Artificial Intelligence in Biobanking

This insightful collection highlights the ethical, legal, and societal issues associated with the increasing role played by artificial intelligence (AI) in medical biobanks, a key research resource in the global study of disease prevention and improving individual care.

Although AI has the potential to speed up health research, the book considers numerous questions that the technology poses, from the building of trust to the prevention of harm to individuals, vulnerable groups, or entire populations. Examining the tension between scientific progress and safeguarding of individual rights, and covering key issues such as accountability, data bias, transparency, and liability, the book considers the legal landscape in which biobanks operate, and what layers of governance are required to oversee such an important resource in a fluid technological age.

A timely volume that brings together scholars and experts from social sciences, ethics, and law, this important book will interest researchers and professionals in Biomedicine, Law, and the broader Health Sciences.

Michaela Th. Mayrhofer is a political scientist, historian, and sociologist specializing in the governance of life sciences. She is founder of Papillon Pathways e.U. and the lead of ELSI at the Medical University of Innsbruck, Institute of Human Genetics, both Austria. Michaela has served as Head of ELSI Services & Research Department (2019–2025) and Co-Interim Director General (2020) at BBMRI-ERIC, Austria. She is a review editor for *Frontiers in Digital Health*, as well as an independent ethics advisor to several consortia and organizations, and a research ethics expert for the European Commission.

Santa Slokenberga is an associate professor in medical law and a senior lecturer in administrative law at the Faculty of Law, Uppsala University, Sweden. She is a board member of the Nordic Permed Law Network and the European Association of Heath Law, as well as an independent ethics advisor to several consortia and organizations.

Signe Mežinska is a bioethicist and sociologist specializing in biomedical research ethics and medical ethics. She leads a research group at the University of Latvia, focusing on ethical challenges in the development and application of biotechnologies and the protection of patients' and research participants' rights. Signe has served as an ethics expert for international organizations, including UNESCO's International Bioethics Committee, WHO, and the European Commission.

Artificial Intelligence in Biobanking

Ethical, Legal and Societal Challenges

Edited by Michaela Th. Mayrhofer, Santa Slokenberga and Signe Mežinska

Routledge
Taylor & Francis Group
LONDON AND NEW YORK

First published 2026
by Routledge
4 Park Square, Milton Park, Abingdon, Oxon OX14 4RN

and by Routledge
605 Third Avenue, New York, NY 10158

Routledge is an imprint of the Taylor & Francis Group, an informa business

British Library Cataloguing-in-Publication Data
A catalogue record for this book is available from the British Library

ISBN: 978-1-032-61992-7 (hbk)
ISBN: 978-1-032-63449-4 (pbk)
ISBN: 978-1-032-63448-7 (ebk)

DOI: 10.4324/9781032634487

Typeset in Times New Roman
by Apex CoVantage, LLC

Contents

Contributors

Mónica Cano Abadía, ELSI Services & Research Department, BBMRI-ERIC, Graz, Austria

Kaya Akyüz, ELSI Services & Research Department, BBMRI-ERIC, Graz, Austria

Leslie Salgado Arzuaga, Communication, Media and Film Department, University of Calgary, Calgary, Canada

Elodie Caboux, International Agency for Research on Cancer, World Health Organization, Lyon, France

Katharina Ó Cathaoir, Faculty of Law, University of Copenhagen, Copenhagen, Denmark

Io Hong Cheong, International School of Public Health, Shanghai Jiao Tong University, Shanghai, China

Anna Clareborn, Department of Immunology, Genetics and Pathology, Uppsala University, Uppsala, Sweden

Valentina Colcelli, Institute IFAC, National Research Council (Italy), Florence, Italy

Ilaria Colussi, ELSI Services & Research Department, BBMRI-ERIC, Graz, Austria

Patrick Courtney, International tec-connection, Konstanz, Germany

Melanie Goisauf, ELSI Services & Research Department, BBMRI-ERIC, Graz, Austria

Zisis Kozlakidis, International School of Public Health, Université Libre de Bruxelles, Brussels, Belgium

Dorota Krekora-Zając, Department of Civil and Commercial Law, University of Białystok, Białystok, Poland

Michaela Th. Mayrhofer, Institute of Human Genetics, Medical University of Innsbruck and Papillon Pathways e.U, Austria

Signe Mežinska, Institute of Clinical and Preventive Medicine, University of Latvia, Riga, Latvia

Jakub Pawlikowski, Department of Humanities and Social Medicine, Medical University of Lublin, Poland

Jarosław Sak, Department of Humanities and Social Medicine, Medical University of Lublin, Poland

Santa Slokenberga, Faculty of Law, Uppsala University, Uppsala, Sweden

Peaking in the Black Box of AI and Its Many Layers

Michaela Th. Mayrhofer, Santa Slokenberga, and Signe Mežinska

1 Introduction

Biobanks are invaluable resources for advancing scientific knowledge, understanding diseases, developing preventive measures, and creating targeted treatments. In health research, biobanks store human biological materials, such as tissues or cells, along with associated participant data, operating at the intersection of healthcare and health research. They serve as essential infrastructures by contributing to the reduction of the global disease burden, while improving healthcare services and personalized medicine. As biobanking practices continue to evolve, emerging technologies – particularly artificial intelligence (AI), machine learning (ML), and big data analytics – are revolutionizing how biobank data is utilized, managed, and shared. These technologies offer new opportunities for accelerating scientific discovery, but they also introduce complex ethical, legal, and societal challenges.

The integration of AI – understood in a broad sense – into biobanking is transforming not only biomedical research but also data governance and decision-making processes. AI-driven analytics can identify intricate patterns in health, genetic, clinical, and environmental data, leading to more precise disease predictions, earlier interventions, and more effective treatments. However, this increasing reliance on AI and big data necessitates a careful examination of the implications for individual rights, societal values, and research governance frameworks. As AI systems continuously refine their algorithms and require increasing amounts of datasets, new tensions arise regarding privacy, data security, informed consent, and equitable access to scientific advancements.

While AI has the potential to enhance health research and improve healthcare delivery, these advancements must be accompanied by a robust consideration of its potential risks. Without appropriate safeguards, AI could lead to harmful medical decisions and incorrect predictions, inadvertently perpetuate biases, disproportionately affect vulnerable populations, or challenge fundamental human rights. Given these risks and the blurred boundaries between health research and healthcare applications, biobanks must navigate the complex terrain of AI's ethical, legal, and societal implications (ELSI) to ensure responsible and fair use of data.

In the absence of uniform legal and ethical frameworks, it is a rather challenging task. The field of AI alone has been subject to legislative actions and ethical

DOI: 10.4324/9781032634487-1

guidelines, such as by the EU and treaty-making by the Council of Europe, to name the central instruments. At the same time, these coexist with legal acts shaping such surrounding central questions as data protection and privacy protection. Additionally, ethical frameworks play a significant role in shaping research ethics review procedures and protecting research participants. Biobankers, researchers, as well as research ethics committees have a particularly challenging task navigating the complexity of the field, which is simultaneously in constant advancement, against the rather still frameworks, balancing the need for scientific advances, while ensuring respect toward individuals and safeguarding their rights.

2 Scope and Objectives of the Edited Volume

This edited volume emerged from a call for abstracts issued in the autumn of 2021 by BBMRI-ERIC. It was followed by an online workshop titled "AI and Biobanking: Ethical, Legal, and Societal Issues of AI in Biobanking" in early spring 2022. This workshop brought together experts from diverse disciplines to explore the ethical, legal, and societal challenges posed by the integration of AI into biobank research and data governance.

Building on the discussions and insights from the workshop, contributors began drafting and refining their chapters, engaging in multiple rounds of revisions. The ongoing legislative developments, particularly the aim of the European Union (EU) to establish a common regulatory and legal framework for AI, most notably the AI Act, played a significant role in shaping these contributions, prompting authors to reassess their analyses in light of emerging regulatory frameworks and evolving debates surrounding AI governance in health research. This iterative process ensured that the volume remains not only academically rigorous but also highly relevant to current policy discussions and future applications of AI in biobanking.

Ultimately, this edited volume brings together leading scholars and experts from the fields of social sciences, ethics, law, philosophy, political science, and biobanking to examine the multifaceted relationship between AI and biobank research. The book specifically explores the mutual impact of ELSI on AI applications in biobanking, as well as the influence of AI on ethical and legal frameworks, such as the AI Act, governing biobank research.

3 Key Questions Addressed in the Volume

By engaging with key interdisciplinary perspectives, this edited volume aims to provide a comprehensive analysis of the evolving landscape of AI-driven biobanking and offer insights into best practices for responsible innovation, especially by addressing several pressing questions at the intersection of AI and biobanking, including:

- *Digitalization and data governance*: What challenges does the digital transformation of biobanking introduce, and what strategies should be adopted to address them?

- *Ethical and human rights considerations:* How do AI systems and big data analytics in biobanking challenge established ethical principles, human rights, and societal values?
- *Scientific progress versus individual and societal interests*: How can we balance the right to benefit from scientific advancements with the need to protect individual autonomy, privacy, and public interests?
- *Regulatory gaps and legal frameworks*: In a globalized world with varying legal standards, how can we manage AI's growing appetite for data while ensuring legal and ethical compliance?
- *Emerging policies and regulations*: How might initiatives such as the AI Act and the European Health Data Space (EHDS) Regulation shape the future of AI applications in biobanking?
- *Need for tailored safeguards:* What new or enhanced safeguards are required to address the specific risks associated with AI-driven biobank research?
- *Research ethics committees and AI oversight*: Do existing research ethics committees have the expertise and capacity to effectively assess AI-based research proposals involving biobanks?
- *Bias and fairness in AI systems*: Can AI fulfil its promise of improving biobank research while mitigating biases in algorithm design and implementation?
- *New vulnerabilities introduced by AI:* Does AI create unforeseen vulnerabilities for individuals, communities, or entire populations?
- *Public engagement and trust-building:* How can societal engagement and patient involvement help address ethical concerns, manage expectations, and foster trust in AI-driven biobank research?

By addressing these questions, this book aims to contribute to ongoing discussions on responsible and trustworthy AI development, the ethical governance of biobank research, and the need for inclusive and transparent decision-making processes in the digital age. As biobanking continues to evolve alongside rapid advancements in AI and big data analytics, a collaborative and interdisciplinary approach will be essential in shaping ethical, legal, and policy frameworks that prioritize both scientific progress and the protection of fundamental rights.

Part 1: Legal Frameworks for AI sets the foundation for this book by discussing the legal frameworks for AI, especially in the EU. In the chapter "The Approach of EU Law to Tackling Statistical Bias in Medicine", Katharina Ó Cathaoir assesses the extent to which the EU legal framework governs the collection of health data and its later use to the development of AI systems, especially in relation to the prevention of statistical biases. Dorota Krekora-Zając discusses in the chapter "Liability for Damages, AI, and Machine Learning for Digital Pathology as New Challenges for Biobanks" how traditional models of liability are challenged, especially through the creation and use of large datasets.

Part 2: Between Risk and Trust expands on aspects of risk and trust. Most notably, concepts of trustworthiness are examined in relation to AI systems and their relevance for biobank management. In the chapter "Unique by Design – the Need for Individualized Models of Trust in Biobanking AI", Anna Clareborn advocates for built-in trustworthy AI measures that allow taking various stakeholder viewpoints into account. Ilaria Colussi suggests considering the precautionary principle, responsible research, and the innovation principle as common ground for inclusive risk management of AI, as highlighted in the chapter "A Possible Risk Governance Approach for AI in Health Research and Biobanking". The authors Kaya Akyüz, Mónica Cano Abadía, and Melanie Goisauf take a critical look at understanding hopes and hypes, as well as risks and opportunities in the chapter "Artificial Intelligence, Sex, and Gender: Hypes, Hopes, and Potential Risks in Biobanking". This is done by using sex and gender as situated data categories to unpack data practices and potential bias.

Part 3: Communicating the Unknown focuses on communication and transparency. The chapter "Fairness of an AI System in the Case of a Biobank of Images and Imaging Biomarkers" by Valentina Colcelli analyzes the legal and ethical dimensions of fairness in the context of imaging biobanks, especially regarding the appropriate representation of patient groups in training datasets. Jarosław Sak and Jakub Pawlikowski suggest that AI can be useful for effective communication with research participants at various stages of biobanking, such as data collection, obtaining informed consent, or providing tailored information on research results. Their chapter is titled "Artificial Intelligence and Communication With Research Participants in the Process of Biobanking Human Biological Material". In the chapter "Tackling AI Transparency Concerns in Biomedical Research: Bringing a Communication-participatory Approach to the Conversation", Leslie Salgado Arzuaga discusses how embracing communicative-participatory approaches can enhance transparency and explainability in the adoption of AI tools.

Part 4: Embracing the Potentiality elaborates on social acceptance, ethical oversight, and the potentiality of AI systems. In the chapter "Social Acceptance of Artificial Intelligence in Biobanking", the authors Elodie Caboux, Patrick Courtney, Io Hong Cheong, and Zisis Kozlakidis identify three dimensions of context-dependent social acceptance: sociopolitical, community, and market acceptance. In relation to biobanking, they criticize the often-overlooked fact that biobanking not only supports research but also serves as a site where research is conducted, which significantly impacts social acceptance. The chapter "Biobank-Based Research Employing AI Techniques: Challenges for Research Ethics Committees" by Signe Mežinska recommends the inclusion of interdisciplinary experts as members of research ethics committees (RECs), capacity training for existing members, and a constant dialogue with researchers to remain current on scientific progress. Michaela Th. Mayrhofer in the chapter "Powerful AI: Between Accountability and Potentiality" examines the strategic and practical dimensions of potentiality ascribed to AI and its relevance for biobanking and embeddedness in social value systems.

4 Acknowledgments

Several chapters benefited from national or EU funding. The chapter by Dorota Krekora-Zając has received funding from the National Science Center Poland

"Agreements for biobanking of human biological samples for scientific purposes" 2016/23/D/HS5/00411. The chapter by Ilaria Colussi is the result of reflections and dialogue conducted within two European projects: HealthyCloud, which has received funding from the European Union's Horizon 2020 research and innovation program under grant agreement No 965345, and Bigpicture under grant agreement No 945358. Valentina Colcelli's contribution has benefited from funding from the European Union's Horizon 2020 research and innovation program under grant agreement No 952159 (ProCAncer-I). Activities conducted under EOSC-Life, a project that has received funding from European Union's H2020 program (grant agreement No 824087), contributed to the writing of the chapter by Kaya Akyüz, Mónica Cano Abadía, and Melanie Goisauf. In addition, several chapters benefited from exchanges with colleagues at conferences, workshops or projects, which is acknowledged in the individual chapters.

We are particularly grateful for having received contributions, in varying amounts, which helped to make the open-access publication possible. This includes funding from the European Union's Horizon 2020 research and innovation program under grant agreements No 952103 (EuCanImage) and No 101016775 (Intervene), as well as the Innovative Medicines Initiative 2 Joint Undertaking under grant agreement No 945358 (Bigpicture). The largest share was covered by BBMRI-ERIC. Moreover, several authors of this volume are affiliated with BBMRI-ERIC, its National Nodes, or the ELSI department and extended network in various capacities. They have benefited from an interdisciplinary dialogue that not only brings together ELSI experts from various disciplines but also allows for continuous exchange with quality managers, pathologists, geneticists, computer scientists, clinic-based biobankers, etc. As co-editors, we thank Amy Thomson and George Russell from Public Health and Allied Health | Routledge Books | Taylor & Francis Group for their excellent editorial support and patience throughout, including the fantastic copy editing team, most notably Bhuvaneshwari Sivakumar. Finally, on behalf of all contributors, we extend our sincere gratitude to Verena Borecký, for her invaluable efforts in herding cats and coordinating the editing process to ensure its submission.

Part 1

Legal Frameworks for AI

1 The Approach of EU Law to Tackling Statistical Bias in Medicine

Katharina Ó Cathaoir

1 Introduction

Academic scholarship increasingly acknowledges that medicine can be biased against groups of patients, impacting their health outcomes (Kadambi 2021; Goodwin and Duke 2012). Think of an app intended to identify skin cancer that proves inaccurate on darker skin tones, having only been tested on a dataset of light-skinned patients (Wen et al. 2022). Or pulse oximeters which were shown to be less reliable on dark skin than light skin during the COVID-19 pandemic (Tobin and Jubran 2022). Meanwhile, sex bias has been identified in fields such as cardiology, where algorithms for detecting risk of heart attacks tested on male datasets may be less accurate when used to diagnose women (Straw et al. 2024). Social biases also feed in; in the US, an algorithm was biased in favor of white patients because healthcare costs were used as a proxy for health needs without accounting for less money being spent on equally sick black patients (Obermeyer et al. 2019).

One cause of statistical bias is data gaps, that is, datasets that fail to reflect population diversity and contain limited data on certain patient groups. Data gaps can result from nonrepresentative clinical trials or population biobanks that fail to collect diverse samples. Within biobanking for genomic research, European samples dominate, for example (Popejoy and Fullerton 2016). Black patients are underrepresented in clinical trials and also more likely than white patients to be excluded from trials (Alegria et al. 2021). Some gaps have their roots in historical exclusionary policies. For example, researchers excluded women from clinical trials due to concerns that hormonal fluctuations could affect the reliability of clinical trials results, as well as potential risks to the fetus (Steinberg et al. 2021; Brazil 2020). Despite having on average poorer health outcomes, researchers often exclude people with cognitive disabilities from health research, meaning that certain conditions go understudied (Shepherd 2020). Older people may have other comorbidities, leading to their exclusion from clinical trials (Stephenson et al. 2012) and off-label medication use due to differences in metabolism (Ren and Zajicek 2015).

Without representative datasets, there is a risk that artificial intelligence (AI) will extend data gaps and resulting biases. Norori et al. (2021) categorize biases as statistical (i.e., due to nonrepresentative datasets) and social (i.e., driven by population-level inequality). If healthcare is to rely increasingly on AI, one response to the risk of bias

DOI: 10.4324/9781032634487-3

is to gather more representative health data, upon which algorithms can be trained to develop diagnostic tools and treatments that are fit for purpose. Representative biobanks are thereby important for the future development of accurate and equitable healthcare technologies. Biobanks are crucial as biological samples are important data sources from which genetic and other data can be extracted, stored, and later processed for multiple uses (Knoppers and Zawati 2013). The data can then be combined with other types of health or demographic data to develop algorithms.

The aim of this chapter is to assess the extent to which the EU legal framework promotes gathering and using representative health datasets to prevent and remediate statistical biases in the development of AI systems for healthcare purposes. Legislation governing the processing of health data and conditions for its use in developing AI devices and treatments is analyzed. The chapter focuses on EU law because there are numerous new and emerging legal acts that harmonize Member State law in the field.

The next section describes the methods used in this chapter. In Section 3, the selected legal acts are introduced. Section 4 outlines the extent to which seven selected concepts are provided for in the relevant regulations. Subsequently, the findings of this analysis are discussed in Section 5, followed by the chapter's conclusions in Section 6.

2 Methods

The chapter maps and compares provisions governing gathering of health data and those addressing bias in AI systems prior and post market authorization. The author chose the following acts based on an assessment of their relevance to the research question: the General Data Protection Regulation (GDPR, Regulation (EU) 2016/679), the Clinical Trials Regulation (CTR, Regulation (EU) 536/2014), the Medical Devices Regulation (MDR), the In Vitro Diagnostic Medical Devices Regulation (IVDMDR), the Data Governance Act (DGA), the European Health Data Space (EHDS), and the AI Act (AIA, Regulation (EU) 2024/1689).

The chapter does not offer a detailed analysis of the relevant acts nor accompanying case law but focuses on analyzing relevant legislative provisions. The analysis is supplemented by the EU Charter, as EU law must be interpreted in light thereof (Article 51(1)). However, the Charter does not provide the basis for an individual entitlement to a particular service or good (O'Cinneide 2014, 736) nor does it require the EU legislature to adopt any specific measure or confer a subjective right on an individual.[1]

To answer the research question, provisions containing the following terms (or proxies) are analyzed: data collection, data quality (proxy: representative), non-discrimination, sex (proxies: gender, woman, and man), age (proxies: older, elderly, and child), disability (proxy: capacity), ethnicity, and race. These groups were chosen because academic research recognizes them as vulnerable to bias in the medical field, and they are covered by the legal protections of EU non-discrimination law.

Yet, the inadequacy of such categorizations is acknowledged. Sex, gender, race, ethnicity, and childhood/old age are social constructs. There is no biological basis for race or ethnicity. Racial inequality is instead the main predictor of health inequalities (Cerdeña et al. 2020). In other words, "race is real, it matters but it is not genetic" (Yu et al. 2009). While algorithms have been shown to disadvantage racial and ethnic groups, this discrimination is not based on biological factors (Eubanks 2017). The author recognizes that categories like socioeconomic position or rural/urban divide may also be powerful in shaping health inequalities (Project Hope 2018). Yet, as these categories are currently not comprehensively protected under EU discrimination law, they are not studied in this chapter.

Furthermore, the manner in which categories are constructed also has implications for the types of data collected and how it is structured. For example, binary categories of man/woman fail to capture implications for trans or non-binary people (see chapter "Artificial Intelligence, Sex, and Gender: Hypes, Hopes, and Potential Risks in Biobanking" by Kaya Akyüz, Mónica Cano Abadía, and Melanie Goisauf). There is no definition or scientific categorization of racial groups and the terms white, black, Asian, etc. are not harmonized (Yearby 2021). It is also important to note that these categories may intersect, that is, older women are underrepresented in clinical trials related to heart disease (Vitale et al. 2017).

3 EU Legal Acts

The relevant legal acts are briefly introduced as follows in chronological order. Due to space constraints, the regulations are not analyzed in detail.

3.1 The Clinical Trials Regulation

The Clinical Trials Regulation (CTR) harmonizes the authorization procedure for clinical trials in the EU. The CTR is a health-specific instrument with a joint legal basis under Articles 114 and 168(4)(c) TFEU like the MDR and IVDR (see Section 3.3), whereas the other acts described in the chapter are of general application.

To briefly summarize the Regulation, it harmonizes supervision and assessment of clinical trials in the EU. Following Article 4 CTR, clinical trials must be subject to scientific and ethical review. The Regulation establishes several new mechanisms, such as a single EU portal for submission of data and information relating to clinical trials (Article 80) and a publicly accessible EU database to facilitate information sharing on clinical trials (Article 81(2)). These measures aim to boost transparency. Several provisions of the CTR are considered relevant to prevent statistical bias. These aspects relate to empowerment of participants through information, sharing of information about clinical trials, and prompts to organizers of clinical trials to include certain groups. The Regulation also aims to protect vulnerable groups, such as pregnant or breastfeeding women, minors, and persons without capacity, by increasing the requirements for their inclusion in clinical trials. These provisions are discussed in Section 4.

3.2 The General Data Protection Regulation

The GDPR aims to harmonize data protection in the EU to facilitate the free movement of data and to protect data subjects' fundamental rights (recitals 1, 2, and 6). Although the Regulation is not a health-focused treaty, the text acknowledges the importance of scientific research for creating knowledge of health conditions and diseases (recital 157). The Regulation presents barriers to the gathering of representative datasets. The prohibition of processing sensitive data is the rule and their processing is the exception. Special provisions apply to processing of children's data, though not other groups. The Regulation simultaneously seeks to empower data subjects through transparency, information, and rights (Articles 12–23). The ethos of the Regulation is that through information, individuals can exercise their data protection rights and protect their personal data (Forcier et al. 2019).

3.3 The Medical Devices Regulation and In Vitro Diagnostic Medical Device Regulation

The Medical Devices Regulation (MDR) establishes a regulatory framework for placing medical devices on the market or putting them into service. It adopts a risk-based approach, whereby devices are classified based on their intended purpose (Annex VII, Article 3.1) into classes I, IIa, IIb, and III (Article 51(1)). Devices with the lowest risk can be placed on the market based on self-assessment by the manufacturer, whereas notified bodies conduct conformity assessments for higher risk devices prior to them being placed on the market (Article 52(1)). Clinical evaluation is required regardless of the risk level attributed to the device (Article 61). The Regulation provides for an EU declaration of conformity (Article 19), CE marking of conformity (Article 20), a Unique Device Identification System to facilitate traceability and identification (Article 27), a UDI database (Article 28), and a European database on medical devices (Article 33). The Commission must set up an electronic system on clinical investigations, which is interoperable with the EU database on clinical trials (MDR, Article 73). Post-market surveillance is provided for under Article 83.

Meanwhile, the IVDR 2017 classifies devices from A (low risk) to D (high risk) (Article 47(1)). Such devices are subject to safety and performance requirements prior to being placed on the market (Article 5(2)). Devices must normally bear a CE marking (Article 18(1)) and be registered with a UDI-DI (Article 26(2)). The manufacturer must provide a summary of safety and performance in the case of high-risk devices (Article 29). Manufacturers must carry out a conformity assessment prior to placing on the market (Article 48(1)). The form of assessment depends on the classification of the device, which is laid out in Annex VIII. Devices are also subject to post-market surveillance, for which the manufacturer is responsible (Article 78). Under the MDR and IVDR, health institutions, such as hospitals, may manufacture, modify, and use devices in-house and be exempt from the Regulation (apart from Annex I), where the "specific needs of target patient groups" cannot be met at the appropriate level of performance by an equivalent device available on the market (Article 5(5) MDR and Article 5 IVDR). Both regulations draw on the language of empowerment, whereby information should be made available to the public (see, e.g., MDR, Article 33). The same groups mentioned in the CTR are also given special protections, see further Section 4.

3.4 Data Governance Act

The DGA seeks to increase trust by establishing mechanisms to facilitate making data available for reuse. The Act does not set out the conditions for reuse but requires that competent public sector bodies make *publicly available* the conditions for allowing data reuse (Article 5(1)). Member States must designate a competent body to assist public sector bodies that grant or refuse data access (or to grant access if empowered to do so) (Article 7). States must furthermore establish single information points to facilitate information regarding conditions for access (Article 8).

Furthermore, the DGA introduces data intermediation services that should support and promote voluntary data sharing between, inter alia, private sector actors (recital 27). The DGA sets notification requirements (Article 11), conditions for (Article 12), and monitoring of (Article 13) such entities. Member States must establish a competent authority responsible for a public national register of data altruism organizations (Article 23(1)). Furthermore, the Commission is required to establish a European data altruism consent form (Article 25) and a European Data Innovation Board (an expert group) (Article 29(1)). The form will enable data subjects to consent to their data being reused based on altruism (i.e., voluntarily and without compensation).

3.5 Artificial Intelligence Act

The AIA lays down harmonized rules for placing on the market, putting into service, and use of AI systems. Like the MDR/IVDR, the AIA adopts a risk-based approach. Certain AI systems are banned (Article 5), high risk AI systems (Article 6) are subject to detailed requirements (Chapter 2), and transparency obligations apply to low-risk AI systems (Article 50).

In the field of healthcare, an AI system will be classified as high risk if it is a safety component or product covered by, inter alia, a medical device subject to a conformity assessment under the MDR (Article 6(1)(a),(b)). Following Article 43(3), the requirements of the AIA shall be added to the conformity assessment already being carried out by notified bodies under, inter alia, the MDR. Annex III identifies the following as high-risk AI applications in healthcare: decisions by public authorities to determine eligibility for healthcare services (Annex III, 1(5)(a)), AI systems intended to be used for risk assessment and pricing in health or life insurance (Annex III, 5(c)), and AI systems to evaluate and classify emergency calls or to be used in dispatching and prioritizing first response, as well as emergency patient triage systems Annex III, 5(d). Such systems will not be subject to third-party conformity assessments involving notified bodies but instead internal control. Furthermore, a system will not be classified as high risk if "it does not pose a significant risk of harm to the health, safety, or fundamental rights of natural persons, including by not materially influencing the outcome of decision making". This could be by, inter alia, performing narrow procedural tasks, or detecting patterns or preparatory tasks, as long as profiling is not performed (Article 5(3)).

As described in Section 4.2, the AIA puts data quality requirements in place. It also includes human rights protections for specific groups. However, it should be noted that the AIA does not apply to AI systems whose sole purpose is scientific research and development (Article 2(6)). Depending on how broadly this exception is interpreted, it could mean that a large number of healthcare applications used in-house in, inter alia, university hospitals are never subject to the data quality requirements.

3.6 European Health Data Space

The aim of the EHDS is to improve "access to and control" over health data in the context of healthcare and access to quality health data for secondary use, for, inter alia, research purposes. Chapter IV of the EHDS requires health data holders to make information on the health data they hold and the data themselves available to health data access bodies (HDABs) for secondary use. Member States must designate one or more HDABs responsible for granting health data access upon request. HDABs may only grant access for secondary use where processing is necessary for specific purposes, including scientific research, which can include development and innovation activities, and training, testing, and evaluating algorithms (Article 53). Health data users must normally make the results or outputs available within 18 months of the completion of the processing (Article 61(4)). These provisions can help to support the gathering of representative datasets, as further described in the next section.

4 Analysis of Qualitative Criteria

As noted earlier, this chapter interprets effective prevention of statistical bias to require gathering representative data, as well as implementing processes to mitigate biases when developing and authorizing AI systems. The following sections assess how the legal instruments surveyed directly or indirectly address seven aspects of this research question: gathering health data, data quality, non-discrimination, sex/gender, age, disability, and ethnicity/race (see further, Section 2).

4.1 Gathering Health Data

The GDPR governs the processing of health data, including their collection, storage, and sharing. Certain provisions create the impression that the Regulation seeks to limit data processing, which can create challenges for health research (Peloquin et al. 2020). These include the GDPR's principles, laid out in Article 5, such as, the purpose limitation and data minimization.

Notably, following Article 6, processing of personal data requires a legal basis, while Article 9(1) prohibits processing of data concerning an identifiable person's health and genetic data unless an exception (legal ground) applies, such as processing based on explicit consent, substantial public interest, or scientific research purposes (Article 9(2)(a), (g), (j)). While consent has typically been understood as central to health research, the Regulation imposes strict conditions, which may render consent an inappropriate basis for gathering data in healthcare or research settings due to inherent power imbalances (EDPB 2019). However, other legal bases are available which may be more research-friendly, such as performance of a task in the public interest. The European Data Protection Board (EDPB) has clarified that personal data can be further processed for scientific research purposes without requiring a new legal basis in line with Article 5(1)(b) GDPR, provided adequate safeguards are respected (EDPB 2019, para 20). Processing for research purposes should provide "appropriate safeguards" for the data subject, such as pseudonymization, encryption, and nondisclosure agreements (EDPB 2020, para 51). Following Article 89, Member States can even impose specifications and derogations with regard to rights to access, rectification, restriction, and objection, which may facilitate the processing of data for scientific research purposes (Comandè and Schneider 2022). Still, Article 89 has been implemented differently in the various Member States, and often requires balancing the interests of the data subject and the controller (Staunton et al. 2019; Molnár-Gábor et al. 2022). Furthermore, Member States can impose limitations on processing of genetic or health data following Article 9(4).

The GDPR aims to promote the free flow of data within the Union, which may have the side effect of increasing the regulatory burden in relation to multi-site research with non-EU countries (Hallinan et al. 2021; Glasdam et al. 2024). Transfers of personal data to a third country may occur based on an adequacy decision of the Commission or subject to appropriate safeguards (Articles 45 and 46). Should this not be in place, the transfer requires a basis in Article 49, such as explicit consent or important reasons in the public interest, accompanied by adequate safeguards.

In response to fragmentation caused by the GDPR, the DGA and EHDS seek to facilitate secondary use of existing datasets. The EHDS provides a legal basis under Article 9(2) and the necessary safeguards for health data holders to make personal electronic health data available to HDABs (EHDS, recital 52). It also places a legal obligation on health data holders to make the data available to the HDAB. The applicant must however demonstrate a valid legal basis under Article 6(1) for processing the health data (Article 67(4)). Furthermore, the EHDS appears to limit the flexibility allowed to Member States under Article 9(4) GDPR. It remains to be seen how the latter provision will interact with Article 51(4) EHDS, which allows Member States to introduce "stricter measures and additional safeguards" for secondary use of genetic, epigenomic, and genomic data, data from wellness apps, health data from biobanks and associated databases, and other human molecular data such as proteomic, transcriptomic, metabolomic, lipidomic, and other omic data.

The DGA harmonizes certain conditions for reuse of public sector data though it does not mandate reuse (Article 1(2)). Public bodies should facilitate reuse of data based on consent (recital 50) or nonpersonal data for objectives in the general interest, such as scientific research (recital 45). Per the DGA, data should normally be anonymized before sharing, although if this would not fit the needs of its reuse, personal data can be shared on-site or remotely through a secure processing environment (recital 15). The risks to the rights and interests of data subjects should be minimal (ibid). Likewise, under the EHDS, HDABs should provide electronic health data in an anonymized format unless the health data user can demonstrate that the processing purpose cannot be achieved with anonymized data. Access must also take place in a secure processing environment.

Meanwhile, the AIA allows that providers of AI systems may exceptionally process special categories of personal data, subject to appropriate safeguards to ensure bias detection and correction following Article 10(2), which is described further in the next section (Article 10(5)). Following Article 59(1), personal data may also be processed in regulatory sandboxes for developing, training, and testing AI systems for safeguarding public interest in, inter alia, public health.

4.2 Data Quality

In contrast to the other regulations, the AIA tackles data quality the most directly. Following Article 10(1), high-risk AI systems which involve training of models must be developed on datasets that meet quality criteria. Datasets must be examined for possible biases that are likely to affect the health and safety of persons, lead to discrimination, or have a negative impact on fundamental rights (Article 10(2)(f)). The availability, quantity, and suitability of datasets must be examined and appropriate measures taken to "detect, prevent and mitigate possible biases" (Article 10(2)(g)). Furthermore, relevant data gaps should be identified and addressed (Article 10(2)(h)). Following Article 10(3),

> Training, validation and testing data sets shall be relevant, sufficiently representative, and to the best extent possible, free of errors and complete in view of the intended purpose. They shall have the appropriate statistical properties, including, where applicable, as regards the persons or groups of persons in relation to whom the high-risk AI system is intended to be used.

Other obligations include that high-risk systems be as "resilient as possible regarding errors, faults or inconsistencies and eliminate as far as possible the risk of possibly biased outputs" (Article 15(4)). Deployers shall ensure to the extent they exercise control over input that the data is "relevant and sufficiently representative in view of the intended purpose of the high-risk AI system" (Article 26(4)).

As part of the obligation to draw up technical documentation for high-risk AI systems, information about the validation and testing data used and their main characteristics shall be provided, as well as discriminatory impacts (Annex IV, 2(g)). Information on the degrees of accuracy for specific persons or groups of persons on which the system is intended to be used will also be provided (Annex IV, 3).

The MDR also contains provisions of relevance to data quality. Member States must ensure that those validating and assessing MDR applications, inter alia, examine "the reliability and robustness of the data generated in the clinical investigation" (Article 71(3)(d)). Furthermore, the clinical environment should be representative of "the intended normal conditions of use of the device in the target patient population" (Annex XV, 2.4). These provisions could be applied to examine and interrogate statistical bias in clinical investigations but the loose formulations do not mandate such an analysis. The IVDR uses less promising language, requiring that an application for assessment includes data showing "the suitability of the device" (Annex X, 2).

In terms of representation, the MDR requires disclosure of the groups included but does not mandate inclusion. For example, the sponsor must provide a summary of the clinical investigation plan, including whether subjects who are minors are included, and the gender and number of subjects (Annex XV, chapter 2, 1.1). If relevant, following the MDR and IVMDR, "information on vulnerable subjects involved such as children, pregnant women, immuno-compromised or, elderly subjects" should be provided (Annex XV, chapter 2, 3.6.3) (Annex XIII, 2.3.2(j)). Special attention must be given to the information needs of specific patient populations (MDR, Article 63 (4)). Following the IVDR, a description of the study population, including inclusion and exclusion criteria must be provided (Annex 2, 6.1.2.6.).

A central aim of the CTR is to allow for a sufficient number of patient populations to take part in clinical trials by involving multiple Member States (recitals 4 and 82). Like the MDR, clinical trials data "should be reliable and robust" (recital 1). The reporting Member State must assess, inter alia, whether the groups of subjects participating in the clinical trial represent the population to be treated, or if not, the explanation and justification provided (Article 6(1)(b)(i)).

The EHDS aims to improve the quality of available health data, including ensuring that they are "as complete as possible" (recital 53). It suggests that data users can enrich datasets, such as by supplementing missing or incomplete data, to improve accuracy or quality of the data (recital 57). Meanwhile, recital 45 of the DGA states that the Regulation should aim to contribute to the emergence of "sufficiently sized" pools of data made available through altruism. This could imply diversity of datasets, although it seems to be a missed opportunity that the concept was not expressly mentioned.

Another goal of the EHDS is to ensure transparency of information concerning datasets. The HDABs must make public a national dataset catalogue with details of the source and nature of electronic health data (Article 57(j)(i)). Health data holders must apply a mandatory data quality label to health datasets made available through HDABs (Article 78(1)). Poor-quality data would still be made available, merely labeled as such (recital 85). Datasets must contain information on, inter alia, accuracy and validity of the data, "representativity of population sampled", and bias examination (Article 78(3)(d)). The Commission will connect national datasets catalogues through an EU Datasets Catalogue (Article 79).

However, the EHDS also makes provision for citizens to "opt out" of secondary use of their data. This is presented as a balance between the need for "exhaustive and representative datasets" with individual autonomy (recital 54). For purposes with a strong link to the public interest, Member States can provide access to datasets even where citizens have exercised their right to opt out.

The GDPR provided data subjects with a range of rights that are strengthened by the EHDS (Ó Cathaoir 2025). While it is beyond the scope of this chapter to explore the rights in detail, it is relevant to note that the EHDS provides individuals with the rights to rectify and insert data into their electronic health record. Alongside digitization of healthcare records due to the requirements of part II of the EHDS, this could contribute to greater availability of quality, accurate healthcare data for research purposes. However, concerns can be raised about the integration of wellness applications into the patient record given associated data quality concerns.

4.3 Non-Discrimination

The AI Act places significant emphasis on non-discrimination to minimize algorithmic bias in datasets and system design. Non-discrimination is a core principle (EP, Article 4(a)), and certain AI systems that violate fundamental rights and non-discrimination should be prohibited (recital 28). When categorizing an AI system as high-risk, its impact on fundamental rights, including non-discrimination, is considered (recital 48). Additionally, the AIA acknowledges the importance of high-quality data to prevent AI systems from becoming sources of discrimination (recital 67).

Fundamental rights assessments were introduced in the final text as a means of combating discrimination. Following Article 27, deployers must conduct a fundamental rights assessment prior to putting high-risk AI systems in use. This should include an assessment of the persons and groups likely to be affected by the use of the AI system, the risks of harm, and measures to be taken (Article 27(1)). Following recital 96, representatives of groups likely to be affected could be involved in conducting such assessments. While the EP had recommended that a summary of the results of the impact assessment should be published (Article 29a(5)), this obligation is regrettably not included in the final text. Even systems that comply with the AIA but present a risk to health or fundamental rights can be required to take measures to ensure the system no longer presents such risk (Article 82).

As with the other EU legal instruments, the AIA sees providing information as an important means of combating risks, such as, of discrimination. AI literacy should be promoted (Articles 4b, 66(f), and 95(2)(c)). Similarly, the EHDS creates obligations for Member States to promote and support digital health literacy (Article 84). Providers of high-risk AI systems shall provide competent authorities with necessary information and documentation to demonstrate conformity (AIA, Article 21). A publicly available and user-friendly EU database for high-risk AI should be established (Article 71). The market surveillance authorities must be granted full access to the training, validation, and testing datasets used for the development of high-risk AI systems for the purposes of effective enforcement (Article 74(12)). National public authorities and bodies that supervise or enforce fundamental rights, including non-discrimination, shall have the power to "request and access documentation" under the Regulation if necessary for effectively fulfilling their mandate (Article 77(1)).

Article 86(1) is one of the only provisions in the AIA that provides entitlements for affected persons, namely a right to a clear and meaningful explanation of the role of the AI system and the main elements of the decision taken in the case of a decision that is taken by a deployer based on the output of a high-risk AI. The decision must produce legal effects or similarly significantly affect the person's health, safety, or fundamental rights. While the EP's text had mentioned "related input data" (Article 68c), this is excluded from the final text. The right would not apply where restricted or exempted under Union or national law. The provision builds on the GDPR's rights in relation to data profiling (Article 4(4)), whereby the data subject has the right to be informed if profiling is being applied to their data (Article 15(1)(h)), and given meaningful information about such processing as well as the significance and consequences thereof (Article 13(2)(f) and Article 14(2)(g)). This includes profiling that consists of automated processing of personal data to evaluate personal aspects, such as prediction related to one's health that has a legal effect (recital 71). Data subjects also have a right to object to profiling (Article 21(1)) and automated decision-making with legal consequences is limited (Article 22(1)). The AIA could have supplemented these entitlements by providing information related to profiling inaccuracies due to biased/unrepresentative datasets but the final text limited this potential.

Discrimination does not feature heavily in the other legal instruments under study. The Data Governance Act and GDPR only mention discrimination in passing as a risk (DGA, recital 5; GDPR, recital 75). Irrespective, EU law prohibits discrimination based on sex, race, and ethnicity in the provision of services such as healthcare.[2] Furthermore, EU regulations must be implemented with respect to the EU Charter, which prohibits discrimination under Article 21 on a wider range of grounds, including age and disability. Thus, when applying and implementing the regulations studied in this chapter, Member States must not directly nor indirectly discriminate provided there is no reasonable justification.

4.4 Sex/Gender

The regulations make modest proposals for boosting gender equality. Several recitals of the AIA recognize that AI systems may lead to discrimination on grounds of gender (recitals 32, 48, and 54). The EP's text had included mentions of gender equality (EP, Article 4a(1)(e); Article 69(2)(c)) but these were not incorporated in the final text. Recital 2 of the Data Governance Act notes that special attention must be paid to, inter alia, boosting the participation of women in the data economy. Under the AIA, codes of conduct shall also reflect the need and relevance of diverse development teams, which presumably encompasses gender representation (AIA, Article 95(2)(d)).

The CTR notes that unless otherwise justified in the protocol, the subjects participating in a clinical trial should represent the population groups, for example, gender and age groups that are likely to use the medicinal product investigated in the clinical trial (recital 14). The gender of the participants should be mentioned in the application dossier, where relevant (Annex 1, 17(h)), as well as the subject inclusion and exclusion criteria (Annex 1, 17(v)). If a specific gender is excluded or underrepresented, an explanation of the exclusion criteria should be given (Annex 1, 17(y)).

Yet, the goal of sex equality (as mandated in Article 23 of the Charter) may be indirectly undermined by the protective measures contained in the MDR, IVDR, and CTR. Thereunder, special measures apply to clinical investigations/performance studies/clinical trials on pregnant or breastfeeding women. Such studies can only be carried out in limited circumstances and are subject to special safeguards that do not apply to the majority of the population. A clinical investigation may only be carried out on these groups where it has the potential to produce direct benefit outweighing risks and burdens, particular care is taken, and no incentives are offered (Article 66). The CTR contains supplementary provisions that allow clinical trials to be carried out without direct benefit so long as the trial cannot be carried out with "comparable effectiveness" on a different patient group, is capable of benefiting pregnant or breastfeeding women or other women in relation to reproduction or other embryos, fetuses, or children, and poses minimal risk (Article 33). There are thereby greater opportunities to run a clinical trial including pregnant or breastfeeding people than a clinical investigation. The cover letter of the trial should mention if pregnant or breastfeeding women are subjects (Annex 1, 6(a)) and a justification for the inclusion of special population should be given (Annex 1, 17(x)). Specific expertise should be considered when assessing clinical trials on, inter alia, pregnant and breastfeeding women (recital 19).

4.5 Age

Although the rights of children and older people are recognized under Articles 24 and 25, respectively, of the EU Charter, children feature more heavily in the instruments under study than older people.

Under the GDPR, children are identified as a group requiring special protection (recitals 38, 58, and 75). Following Article 8, Member States can set the age of digital consent at between 13 and 16 years, before which, parents/guardians must provide consent on the child's behalf. This can create discrepancy between the age at which a child can consent to medical treatment and data processing. Irrespective of capacity, children are entitled to information in "clear and plain language" (recital 58). Finally, under the obligation to promote public awareness and understanding regarding data processing, children must be given specific attention (Article 57(1)(b)).

The final text of the AIA includes greater reference to age as a category instead of a particular age group, which is an inclusive step (recital 56–58 and 132). Children are given special consideration, with their rights noted in the preamble (recital 28). Providers must give specific consideration as to whether a high-risk AI system is likely to have a negative impact on persons under 18 or other vulnerable groups of people (Article 9(9)). The Parliament's text had moved beyond the Commission's original focus on chronological age and was instead more focused on vulnerability. AI systems that would exploit vulnerabilities of any person, including due to their age, are prohibited (Article 5(1)(b)).

The MDR affords children special protections in recognition of their vulnerabilities (recital 72). The clinical investigation must be intended to investigate treatments for a medical condition that only occurs in minors or must be essential with respect to minors to validate data obtained in clinical investigations on persons able to give informed consent or by other research methods (Article 65(e)). It must relate to a medical condition from which the minor suffers or can only be carried out on minors and there must be scientific grounds for expecting the minor will receive direct benefit (Article 65(f)-(g)). Clinical investigations on minors may only be conducted where informed consent has been obtained from their legally designated representative (Article 65(a)). Minors must also receive information in a way adapted to their age and mental maturity and from investigators or members of the investigating team who are trained or experienced in working with children (Article 65(b)). Their "explicit wish" (provided they are capable of forming an opinion and assessing the information) must be respected (Article 65(c)). The minor must take part in the informed consent procedure in a way adapted to his or her age and mental maturity and if they reach adulthood during the investigation, informed consent must be given to continue participation (Article 65(h)–(i)). These provisions can be suspended in emergency situations, following Article 68 MDR and Article 64 IVDR.

The CTR is more far-reaching in terms of inclusion of age groups. The preamble notes that population groups likely to use the medicine, including age groups, should be represented in the clinical trial, unless otherwise justified (recital 14). The preamble also mentions children's rights under the Charter (recital 83). Furthermore, frail and older people are identified as vulnerable groups, upon whom medicinal products likely to be of significant clinical value should be fully and appropriately studied (recital 15). Minors and older people in residential care should be given special protection (recitals 27 and 35). Specific expertise should be considered when assessing clinical trials involving, inter alia, minors and elderly people (recital 19). Specific consideration must be given to the assessment of an application for authorization of a trial on the basis of pediatric expertise (Article 10(1)). Under the MDR and CTR, Member States can impose additional national measures for persons in residential care institutions, which can have relevance for some older people who are living in, inter alia, care homes (MDR, Article 67 and CTR, Article 34).

The same special rules apply to clinical trials on minors as to clinical investigations under the MDR, with the exception of paragraph (g) whereby a minor can take part in a clinical trial provided

> some benefit for the population represented by the minor concerned and such a clinical trial will pose only minimal risk to, and will impose minimal burden on, the minor concerned in comparison with the standard treatment of the minor's condition.
>
> (Article 32(1)(g))

The same rules also apply to performance studies on minors under the IVDR (Article 61). The measures pertaining to pregnant and breastfeeding women in Annex 1, described earlier, also apply to minors (Annex 1, 6; Annex 1, 17(h); and Annex 1, 17(v)).

Meanwhile, the CTR is supplemented by the Pediatric Regulation, which aims to further the development of medicinal products for human use in order to meet the specific therapeutic needs of the pediatric population, without subjecting the pediatric population to unnecessary clinical or other trials and in compliance with CTR.[3] In a ten-year evaluation, the Commission found that the proportion of clinical trials that include children has increased by 50% in 2007–2016 (from 8.25% to 12.4% of all trials conducted in the EU) and that 260 new medicines for children were authorized between 2007 and 2016. Still, challenges remain, especially in the field of rare diseases (European Commission 2017). Furthermore, a different study found that new medicines authorized for children were often not marketed, thus undermining availability (Lepola et al. 2016). No similar regulation is in place for older people.

4.6 Disability

The rights of people with disabilities are recognized under Article 26 of the EU Charter and the Convention on the Rights of Persons with Disabilities (CRPD), to which the EU is a party.[4] The acts under study make modest references to these rights.

The recitals of the final text of the AIA include greater references to the rights of persons with disabilities than earlier versions. For example, recital 80 references the CRPD and the Union and Member States' obligations to protect persons with disabilities from discrimination and ensure equality. Codes of conduct must in particular assess how AI systems affect vulnerable people, including accessibility for people with disabilities (Article 95(2)(e)). While the Proposal had included an obligation to provide information and notifications in accessible formats (Article 52(3)(b)), this is moved to the recital in the final text.

The EHDS likewise includes a greater emphasis on the rights of persons with disabilities. Electronic health records shall be accessible for vulnerable people, including people with disabilities (Article 4(5)). When providing information on datasets, the accessibility needs of people with disabilities shall be taken into account (recital 84). Particular attention should be given to improving the literacy of people with disabilities and other groups, including the elderly (recital 89). The IVDR underscores that special attention should be paid to the information needs of specific patient populations (Article 59(4); Annex I, Article 20.I(a)). The DGA references the right of integration of people with disabilities and that design for all should be taken into account (recital 46).

The MDR provides special protections for people who are legally incapacitated (recital 72). Clinical investigations on incapacitated subjects who have not given, or have refused to give, informed consent before the onset of their incapacity, can only take place provided firstly, their legally designated representative has given informed consent (Article 64(1)(a)). This form of consent can be criticized for transferring the individual's legal rights contrary to the spirit of the CRPD. The MDR does however require that incapacitated subjects receive information in a way that is adequate in view of their capacity to understand it, and that their explicit wish to refuse participation in, or to withdraw from, the clinical investigation at any time, is respected (Article 64(1)(b), (c)). Still, it is lacking that only the wishes of an incapacitated subject "who is capable of forming an opinion and assessing the information" are determinative (Article 64(1)(c)). As earlier, no incentives may be offered and the clinical investigation must be essential to incapacitated subjects and cannot be instead performed on persons able to consent or by other research methods (Article 64(1)(d), (e)). The subject must suffer from a directly related medical condition and there must be scientific grounds for expecting a direct benefit for the individual (Article 64(1)(f), (g)). Finally, the subject must as far as possible take part in the informed consent procedure (Article 64(2)).

The same special protections apply in the case of clinical trials although it is possible to participate in a clinical trial without direct benefit provided "some benefit" is expected for the population represented by the incapacitated subject when the clinical trial relates directly to the life-threatening or debilitating medical condition from which the subject suffers and provided minimal risk and burden apply (Article 31(1)). Following Article 10(2) CTR, specific expertise should be considered when assessing clinical trials involving incapacitated subjects. The rules determining capacity are left to Member State law (recital 27). More stringent rules can be set at national level prohibiting the conduct of those clinical trials on incapacitated subjects where there are no scientific grounds to expect that participation in the clinical trial will produce a direct benefit to the subject outweighing the risks and burdens involved (Article 31(2)).

4.7 Ethnicity/Race

There are limited provisions expressly mentioning ethnicity and race in the regulations, though discrimination on such grounds is prohibited following Article 21 EU Charter. Under Article 9(1) GDPR, information revealing racial or ethnic origin about an identified or identifiable person is a special category of personal data. There is no requirement to report the race or ethnicity of clinical trial/investigation participants, unlike in the case of gender. The final text of the AIA mentions race and ethnicity as a possible ground for discrimination in its recitals. It is prohibited to place on the market a system that categorizes individual persons based on their biometric data to deduce or infer their, inter alia, race (Article 5(1)(g)).

5 Analysis

The central aim of this chapter was to examine whether and in what ways EU law contributes to gathering representative health data and tackling bias in the development of AI healthcare technologies. Provisions related to seven categories were described separately, although they are likely to intersect. Several patterns emerged that are analyzed in this section.

Firstly, in terms of contributing to gathering representative datasets, the GDPR occupies a central role. As described earlier, the GDPR imposes conditions on researchers who seek to gather representative datasets, which can be experienced as a regulatory burden (Vukovic et al. 2022). Further guidance from the EDPB on the application of the GDPR to scientific research is long-awaited. While the GDPR generally does not contain special provisions for particular groups, it requires a legal basis for processing personal data and a legal ground for processing sensitive data, such as information related to an identifiable or identified person's health (including disability status), sex, ethnicity, or race. Statistical biases may be difficult to detect if data has not been gathered on the subjects' characteristics, thereby rendering such processing necessary (Norori et al. 2021). It is thereby welcomed that Article 10(5) AIA allows the processing of special categories of personal data for bias detection purposes.

The DGA and EHDS aim to contribute to data access and thereby increase the availability of quality research data. The EHDS should promote sharing of data for secondary use, thereby furthering data availability and potentially indirectly mitigating discrimination. The Acts do not expressly address data gaps or contain provisions targeted at increasing gathering or sharing of data pertaining to specific groups. The inclusion of an opt-out mechanism in the EHDS may undermine diversity if a substantial number of citizens exercise this right. It is thereby important that Member States build trust among all groups in society regarding secondary use. Furthermore, the EHDS can be criticized for not fully resolving the issues relating to fragmented legal bases across the EU.

Secondly, the regulations emphasize the importance of providing accessible information, including special requirements for children and persons with disabilities. This can enable individuals to exercise, inter alia, their data protection rights and correct or limit the use of incorrect information. At the same time, it must be asked whether the impressive catalogue of rights contained in the GDPR (and now EHDS) expects too much of the data subjects in terms of knowledge, time, and effort (van Ooijen and Vrabec 2019; van de Waerdt 2020). In relation to medical devices and clinical trials, information should empower individuals to decide on an informed basis whether to participate in investigations or trials. Yet, the information requirements of clinical trials have been criticized in the literature as placing too great a burden on participants. It is necessary to find solutions whereby individuals can exercise their rights, while not being overburdened by technical information (Jenkins et al. 2020). In this regard, it is positive that the AIA and EHDS place increased emphasis on developing digital literacy in society and it is hoped that such initiatives will be adequately resourced.

Thirdly, the AIA, EHDS, CTR, and MDR impose specific positive obligations related to data quality/representativeness. The AIA tackles bias mitigation in the most direct manner of the instruments surveyed. However, it remains to be seen how effective the provisions will prove in practice. For example, what standard of representation is required to comply with AIA's Article 10? How can conditions be put in place that avoid biased AI in healthcare while avoiding standards so far-reaching as to be scientifically implausible? Could a deployer be relieved of their duties simply by labelling a device as, for example, inadequately tested on people with dark skin? The extent to which deployers can identify specific groups as "intended" in order to meet the quality requirements, while avoiding falling foul of the principle of non-discrimination, is in need of clarification. These conditions were ultimately weakened in the final text of the AIA due to feasibility concerns. As with the GDPR, the AIA is not specific to healthcare, and guidance is needed on the application of its provisions to the various sectors, including healthcare, to secure legal certainty.

Likewise, while the MDR and CTR include nudges to develop representative devices/treatments, they are framed broadly, meaning that much is left up to interpretation. There is a risk that the provisions become box-ticking exercises that are not taken seriously and do not produce real diversity. They may simply place the burden on the patient to examine clinical trials data to check whether the trial was representative, without increasing actual representation. Alternatively, for medical devices, the requirements could lead to fragmentation due to differences in how notified bodies interpret them. As Meszaros et al. point out, reviewing large and complex datasets requires specialist knowledge, which notified bodies may lack and may create data processing issues in the case of devices developed outside the EU (Meszaros et al. 2022).

Fourthly, with the exception of the AIA, the regulations barely reference non-discrimination. While the EU Charter applies horizontally, its scope to date in CJEU case law has been criticized (Ward 2018). There is a risk that when deciding applications for authorization, Article 21 will be neglected or alternatively be applied in a manner that leads to fragmentation. Therefore, the instruments' focus on bias and discrimination could have been strengthened through incorporating and clarifying positive obligations when developing medications or devices. While the AIA places discrimination front and center, clarification is needed as to which types of discrimination would be justifiable in healthcare. For example, would an AI system that misinterprets the risk of heart disease among women having only been developed on datasets from men be authorized provided it is labelled as such?

Fifthly, provisions related to specific groups were studied. While discrimination on grounds of ethnicity or race is prohibited under EU law, provisions on ethnicity/race barely feature in the surveyed instruments. It was acknowledged at the outset of this chapter that though race is not biological, gathering self-reported data can be important to identify health disparities and to monitor that devices do not perform in a biased manner. The EU approach can be contrasted with that of the US, whereby the US federal government mandates the inclusion of diverse populations, although the rules have been criticized for failing to benefit people of color (Roberts 2008). Representation in clinical trials and personalized medicine is recognized as a priority through Sec. 907 of the Food and Drug Administration Safety and Innovation Act (FDASIA), NIH Revitalization of Act of 1993, and the "All of US" initiative. However, the success of such initiatives has been questioned (Adashi and Cohen 2023).

Similarly, there is also limited acknowledgment of the rights of older people in the regulations studied, though their right to lead a life of dignity and to participate in cultural life is recognized in Article 25 of the EU Charter. Children feature more heavily in the regulations and have been recognized as a group in need of greater inclusion through, in particular, the pediatric regulation. The AIA and EHDS mark a shift toward focusing on vulnerability instead of particular age groups, which may be a positive step.

The instruments are also modest in terms of imposing positive obligations related to sex/gender. The CTR appears to be the most empowerment-focused instrument, as it calls for the inclusion of those likely to use particular medicines in the clinical trial, unless otherwise justified. Yet, the operative paragraphs of the CTR do not contain positive obligations to include certain groups or persons in clinical trials in order to tackle the data gap described in the introduction. The cover letter to a CTR application should, however, mention the gender of the participants and provide a justification if specific genders are not represented. This aspect may encourage the inclusion of women in clinical trials, although it remains unclear what justification would be accepted to satisfy this provision. There is a risk that this will vary among Member States given that the weight of this provision relies on interpretation by the relevant authority.

Finally, under the CTR and MDR, special justifications are required for involving pregnant or breastfeeding women, children, and persons without capacity in clinical trials or investigations. These provisions should be applied in a proportionate manner that supports participation in line with the individual's wishes, scientific needs, and the rights recognized in the Charter. The regulations do not authorize blanket bans on any group's participation and it is important that they are not interpreted in a disproportionate manner that limits participation based on a perceived notion of protection (Ó Cathaoir 2024).

6 Conclusion

This chapter has assessed the extent to which the EU legal framework governing the collection of health data and its later use for developing AI systems for healthcare purposes accounts for the need to prevent and remediate statistical biases by gathering representative data. Based on this analysis, the relevant regulations include both empowering and protective elements.

The approach of the MDR and CTR is that certain groups, like children and pregnant people, should be protected from misuse in light of historical abuses. There is a risk that such measures may inhibit the collection of representative data for the affected groups. Researchers may avoid including groups like children, pregnant people, and persons with disabilities due to the increased regulatory burden. At the same time, certain aspects go beyond protections, like those in the Clinical Trials Regulation that mention increasing representation. Meanwhile, the Pediatric Regulation could offer a blueprint for other neglected populations, such as older people.

The AIA marks a shift in the EU's approach. It creates positive obligations to prevent and address bias and prohibit discrimination, while the other instruments are framed less prescriptively. These provisions will apply to certain AI medical devices and certain algorithms that have been fed data from clinical trials.

The chapter has remained neutral as to the merits or drawbacks of a protective versus autonomy-focused approach. Furthermore, it has limited itself to a bird's eye view of the EU framework. Given that regulations like the GDPR allow EU Member States a measure of discretion in implementing certain provisions, account must be had of how the Member States have transposed the relevant provisions into law. Furthermore, as many provisions in the CTR and MDR are loosely framed, it is necessary to explore how they are interpreted and applied by the bodies deciding applications for clinical trials or investigations. Such bodies should be adequately supervised and guided at Union level, as envisaged by the different regulations. The IVDR provides, for example, for exchange of ideas and experiences between authorities responsible for notified bodies following Article 44.

The biases outlined in Section 1 are unlikely to be adequately addressed by the emerging legal order described in this chapter. In the case of medical devices, like oximeters, there is no explicit requirement to test devices on, inter alia, a variety of skin colors. Similarly, while there is a requirement to disclose whether women or men have been excluded from clinical trials, it is unclear whether such exclusion would result in a trial not being authorized. It remains to be seen what impact the AIA will have. Would the racially biased healthcare algorithm be approved in the EU or be identified as biased from the outset? Would the algorithm be prohibited as discriminatory or merely labelled as only suitable for certain groups? Finally, it is important to examine how even algorithms developed on representative data perform in real-world conditions (Kostick-Quenet et al. 2022).

Notes

1 See case *Wolfgang Glatzel v Freistaat Bayern.*

2 Council Directive 2000/43/EC of 29 June 2000 implementing the principle of equal treatment between persons irrespective of racial or ethnic origin, OJ L 180, 19.7.2000, pp. 22–6; Directive 2006/54/EC of the European Parliament and of the Council of 5 July 2006 on the implementation of the principle of equal opportunities and equal treatment of men and women in matters of employment and occupation (recast), OJ L 204, 26.7.2006, pp. 23–36.

3 Regulation (EC) No 1901/2006 of the European Parliament and of the Council of 12 December 2006 on medicinal products for pediatric use; Regulation (EC) No 1902/2006, an amending regulation in which changes to the original text were introduced relating to decision procedures for the European Commission.

4 UN General Assembly, Convention on the Rights of Persons with Disabilities, 13 December 2006, A/RES/61/106.

Reference List

Adashi, Eli Y., and I. Glenn Cohen. 2023. "The FDA Initiative to Assure Racial and Ethnic Diversity in Clinical Trials." *The Journal of the American Board of Family Medicine* 36 (2): 366. https://doi.org/10.3122/jabfm.2022.220290R1.

Alegria, Mikhaila, Shawn Sud, Benjamin E. Steinberg, Nan Gai, and Asad Siddiqui. 2021. "Reporting of Participant Race, Sex, and Socioeconomic Status in Randomized Clinical

Trials in General Medical Journals, 2015 vs 2019." *JAMA Network Open* 4 (5): e2111516–e2111516. https://doi.org/10.1001/jamanetworkopen.2021.11516.
Brazil, Rachel. 2020. "Why We Need to Talk about Sex and Clinical Trials." *The Pharmaceutical Journal*, May 28. https://pharmaceutical-journal.com/article/feature/why-we-need-to-talk-about-sex-and-clinical-trials.
Cerdeña, Jessica P., Marie V. Plaisime, and Jennifer Tsai. 2020. "From Race-Based to Race-Conscious Medicine: How Anti-Racist Uprisings Call Us to Act." *The Lancet* 396 (10257). Elsevier: 1125–8. https://doi.org/10.1016/S0140-6736(20)32076-6.
Comandè, Giovanni, and Giulia Schneider. 2022. "Differential Data Protection Regimes in Data-Driven Research: Why the GDPR Is More Research-Friendly Than You Think." *German Law Journal* 23 (4). Cambridge University Press: 559–96. https://doi.org/10.1017/glj.2022.30.
EDPB. 2019. *Opinion 3/2019 Concerning the Questions and Answers on the Interplay between the Clinical Trials Regulation (CTR) and the General Data Protection Regulation (GDPR)*. https://www.edpb.europa.eu/our-work-tools/our-documents/opinion-art-70/opinion-32019-concerning-questions-and-answers_en.
EDPB. 2020. Guidelines 03/2020 on the processing of data concerning health for the purpose of scientific research in the context of the COVID-19 outbreak Adopted on 21 April 2020. https://www.edpb.europa.eu/sites/default/files/files/file1/edpb_guidelines_202003_healthdatascientificresearchcovid19_en.pdf.
Eubanks, Virginia. 2017. *Automating Inequality: How High-Tech Tools Profile, Police, and Punish the Poor*. New York: St. Martin's Press.
European Commission. 2017. *Report from the Commission to the European Parliament and the Council State of Paediatric Medicines in the EU - 10 Years of the EU Paediatric Regulation*. https://www.europarl.europa.eu/RegData/docs_autres_institutions/commission_europeenne/com/2017/0626/COM_COM(2017)0626_EN.pdf.
Forcier, Mélanie Bourassa, Hortense Gallois, Siobhan Mullan, and Yann Joly. 2019. "Integrating Artificial Intelligence into Health Care through Data Access: Can the GDPR Act as a Beacon for Policymakers?" *Journal of Law and the Biosciences* 6 (1): 317–35. https://doi.org/10.1093/jlb/lsz013.
Glasdam, Stinne, Katharina Ó Cathaoir, and Sigrid Stjernswärd. 2024. "Balancing Different Legal and Ethical Requirements in the Construction of Informed Consents in Qualitative International Collaborative Research Across Continents: Reflections from a Scandinavian Perspective." *Journal of Academic Ethics*. https://doi.org/10.1007/s10805-024-09536-5.
Goodwin, Michele, and Naomi Duke. 2012. "Health Law: Cognitive Bias in Medical Decision Making." In *Implicit Racial Bias Across the Law*, edited by Justin D. Levinson and Robert J. Smith. Cambridge: Cambridge University Press.
Hallinan, Dara, Alexander Bernier, Anne Cambon-Thomsen, Francis P. Crawley, Diana Dimitrova, Claudia Bauzer Medeiros, Gustav Nilsonne, Simon Parker, Brian Pickering, and Stéphanie Rennes. 2021. "International Transfers of Personal Data for Health Research Following Schrems II: A Problem in Need of a Solution." *European Journal of Human Genetics* 29 (10): 1502–9. https://doi.org/10.1038/s41431-021-00893-y.
Jenkins, Simon Paul, Melanie J. Calvert, and Heather Draper. 2020. "Potential Research Participants' Use of Information during the Consent Process: A Qualitative Pilot Study of Patients Enrolled in a Clinical Trial." *PLoS ONE* 15 (6). Public Library of Science: e0234388. https://doi.org/10.1371/journal.pone.0234388.
Kadambi, Achuta. 2021. "Achieving Fairness in Medical Devices." *Science* 372 (6537). www.science.org/doi/abs/10.1126/science.abe9195?rss=1.
Knoppers, Bartha, and Ma'n H. Zawati. 2013. "Biobanks." In *Encyclopedia of Applied Ethics*, Volume 1, 246–50. https://doi.org/10.1016/B978-0-12-373932-2.00022-3.
Kostick-Quenet, Kristin M., I. Glenn Cohen, Sara Gerke, Bernard Lo, James Antaki, Faezah Movahedi, Hasna Njah, Lauren Schoen, Jerry E. Estep, and J. S. Blumenthal-Barby. 2022. "Mitigating Racial Bias in Machine Learning." *Journal of Law, Medicine & Ethics* 50 (1): 92–100. https://doi.org/10.1017/jme.2022.13.

Lepola, Pirkko, Allison Needham, Jo Mendum, Peter Sallabank, David Neubauer, and Saskia de Wildt. 2016. "Informed Consent for Paediatric Clinical Trials in Europe." *Archives of Disease in Childhood* 101 (11): 1017–25. https://doi.org/10.1136/archdischild-2015-310001.

Meszaros, Janos, Marcelo Corrales Compagnucci, and Timo Minssen. 2022. "The Interaction of the Medical Device Regulation and the GDPR: Do European Rules on Privacy and Scientific Research Impair the Safety and Performance of AI Medical Devices?" In *The Future of Medical Device Regulation: Innovation and Protection*, edited by Carmel Shachar, Christopher Robertson, I. Glenn Cohen, Timo Minssen, and W. Nicholson Price II. Cambridge: Cambridge University Press. https://doi.org/10.1017/9781108975452.007.

Molnár-Gábor, Fruzsina, Julian Sellner, Sophia Pagil, Santa Slokenberga, Olga Tzortzatou-Nanopoulou, and Katarina Nyström. 2022. "Harmonization after the GDPR? Divergences in the Rules for Genetic and Health Data Sharing in Four Member States and Ways to Overcome Them by EU Measures: Insights from Germany, Greece, Latvia and Sweden." *Precision Medicine in Cancer* 84 (September): 271–83. https://doi.org/10.1016/j.semcancer.2021.12.001.

Norori, Natalia, Qiyang Hu, Florence Marcelle Aellen, Francesca Dalia Faraci, and Athina Tzovara. 2021. "Addressing Bias in Big Data and AI for Health Care: A Call for Open Science." *Patterns* 2 (10): 100347. https://doi.org/10.1016/j.patter.2021.100347.

Obermeyer, Ziad, Brian Powers, Christine Vogeli, and Sendhil Mullainathan. 2019. "Dissecting Racial Bias in an Algorithm Used to Manage the Health of Populations." *Science* 366 (6464). https://doi.org/10.1126/science.aax2342.

Ó Cathaoir, Katharina. 2024. "Deviating from the Norm? The Pregnant Body in Scandinavian Health Law." *German Law Journal* 25 (5): 665–89. https://doi.org/10.1017/glj.2024.11.

Ó Cathaoir, Katharina. 2025. "The EHDS and Electronic Health Records: GDPR+ or Transforming Patients' Rights?" In *Health and Genetic Data: Legal, Governance and Administrative Transformations in the Light of the European Health Data Space*, edited by Santa Slokenberga, Katharina Ó Cathaoir and Mahsa Shabani. Routledge.

O'Cinneide, Colm. 2014. "The Rights of the Elderly." In *The EU Charter of Fundamental Rights: A Commentary*, edited by Steve Peers, Tamara Hervey, Jeff Kenner and Angela Ward. London: Hart Publishing.

Ooijen, I. van, and Helena U. Vrabec. 2019. "Does the GDPR Enhance Consumers' Control Over Personal Data? An Analysis from a Behavioural Perspective." *Journal of Consumer Policy* 42 (1): 91–107. https://doi.org/10.1007/s10603-018-9399-7.

Peloquin, David, Michael DiMaio, Barbara Bierer, and Mark Barnes. 2020. "Disruptive and Avoidable: GDPR Challenges to Secondary Research Uses of Data." *European Journal of Human Genetics* 28 (6): 697–705. https://doi.org/10.1038/s41431-020-0596-x.

Popejoy, Alice B., and Stephanie M. Fullerton. 2016. "Genomics Is Failing on Diversity." *Nature* 538 (7624): 161–4. https://doi.org/10.1038/538161a.

Project HOPE. 2018. "Health, Income, & Poverty: Where We Are & What Could Help." https://doi.org/10.1377/hpb20180817.901935.

Ren, Zhaoxia, and Anne Zajicek. 2015. "Review of the Best Pharmaceuticals for Children Act and the Pediatric Research Equity Act: What Can the Obstetric Community Learn from the Pediatric Experience?" *Seminars in Perinatology* 39 (7): 530–1. https://doi.org/10.1053/j.semperi.2015.08.006.

Roberts, Dorothy E. 2008. "Is Race-Based Medicine Good for Us?: African American Approaches to Race, Biomedicine, and Equality." *The Journal of Law, Medicine & Ethics* 36 (3): 537–45. https://doi.org/10.1111/j.1748-720X.2008.302.x.

Shepherd, Victoria. 2020. "An Under-Represented and Underserved Population in Trials: Methodological, Structural, and Systemic Barriers to the Inclusion of Adults Lacking Capacity to Consent." *Trials* 21 (1): 445. https://doi.org/10.1186/s13063-020-04406-y.

Staunton, Ciara, Santa Slokenberga, and Deborah Mascalzoni. 2019. "The GDPR and the Research Exemption: Considerations on the Necessary Safeguards for Research Biobanks." *European Journal of Human Genetics* 27 (8): 1159–67. https://doi.org/10.1038/s41431-019-0386-5.

Steinberg, Jecca R., Brandon E. Turner, Brannon T. Weeks, Christopher J. Magnani, Bonnie O. Wong, Fatima Rodriguez, Lynn M. Yee, and Mark R. Cullen. 2021. "Analysis of Female Enrollment and Participant Sex by Burden of Disease in US Clinical Trials Between 2000 and 2020." *JAMA Network Open* 4 (6): e2113749–e2113749. https://doi.org/10.1001/jamanetworkopen.2021.13749.

Stephenson, Anne, Geoffrey M. Anderson, and Paula Rochon. 2012. "Off-Label Prescribing in Older People: The Need for Increased Awareness and Caution." *Drugs & Aging* 29 (6): 435–6. https://doi.org/10.2165/11634620-000000000-00000.

Straw, Isabel, Geraint Rees, and Parashkev Nachev. 2024. "Sex-Based Performance Disparities in Machine Learning Algorithms for Cardiac Disease Prediction: Exploratory Study." *Journal of Medical Internet Research* 26: e46936. https://doi.org/10.2196/46936.

Tobin, Martin J., and Amal Jubran. 2022. "Pulse Oximetry, Racial Bias and Statistical Bias." *Annals of Intensive Care* 12 (1): 2. https://doi.org/10.1186/s13613-021-00974-7.

Vitale, Cristiana, Massimo Fini, Ilaria Spoletini, Mitja Lainscak, Petar Seferovic, and Giuseppe MC Rosano. 2017. "Under-Representation of Elderly and Women in Clinical Trials." *International Journal of Cardiology* 232 (April): 216–21. https://doi.org/10.1016/j.ijcard.2017.01.018.

Vukovic, Jakov, Damir Ivankovic, Claudia Habl, and Jelena Dimnjakovic. 2022. "Enablers and Barriers to the Secondary Use of Health Data in Europe: General Data Protection Regulation Perspective." *Archives of Public Health* 80 (1): 115. https://doi.org/10.1186/s13690-022-00866-7.

Waerdt, Peter J. van de. 2020. "Information Asymmetries: Recognizing the Limits of the GDPR on the Data-Driven Market." *Computer Law & Security Review* 38 (September): 105436. https://doi.org/10.1016/j.clsr.2020.105436.

Ward, Angela. 2018. "The Impact of the EU Charter of Fundamental Rights on Anti-Discrimination Law: More a Whimper Than a Bang?" *Cambridge Yearbook of European Legal Studies* 20 (2018): 32–60. https://doi.org/10.1017/cel.2018.11.

Wen, David, Saad M. Khan, Antonio Ji Xu, Hussein Ibrahim, Luke Smith, Jose Caballero, Luis Zepeda, et al. 2022. "Characteristics of Publicly Available Skin Cancer Image Datasets: A Systematic Review." *Lancet Digit Health* 4 (1): e64–e74. https://doi.org/10.1016/S2589-7500(21)00252-1.

Yearby, Ruqaiijah. 2021. "Race Based Medicine, Colorblind Disease: How Racism in Medicine Harms Us All." *The American Journal of Bioethics* 21 (2): 19–27. https://doi.org/10.1080/15265161.2020.1851811.

Yu, Joon-Ho, Sara Goering, and Stephanie M. Fullerton. 2009. "Race-Based Medicine and Justice as Recognition: Exploring the Phenomenon of BiDil." *Cambridge Quarterly of Healthcare Ethics* 18 (1): 57–67. https://doi.org/10.1017/S0963180108090099.

Legal Sources

Case C-356/12 Wolfgang Glatzel v Freistaat Bayern ECLI:EU:C:2014:350 [2014].

Council Directive 2000/43/EC of 29 June 2000 Implementing the Principle of Equal Treatment between Persons Irrespective of Racial or Ethnic Origin [2000] OJ L 180, pp. 22–6.

Directive 2006/54/EC of the European Parliament and of the Council of 5 July 2006 on the Implementation of the Principle of Equal Opportunities and Equal Treatment of Men and Women in Matters of Employment and Occupation (Recast) [2006] OJ L 204, pp. 23–36.

Regulation (EC) 1901/2006 on Medicinal Products for Paediatric Use and Amending Regulation (EEC) No 1768/92, Directive 2001/20/EC, Directive 2001/83/EC and Regulation (EC) No 726/2004 [2006] OJ L 378.

Regulation (EC) 1902/2006 Amending Regulation 1901/2006 on Medicinal Products for Paediatric Use [2006] OJ L 378.

Regulation (EU) 536/2014 on Clinical Trials on Medicinal Products for Human Use [2014] OJ L 158.

Regulation (EU) 2016/679 on the Protection of Personal Data of Natural Persons with Regard to the Processing of Personal Data and on the Free Movement of Such Data (General Data Protection Regulation) [2016] OJ L 119.
Regulation (EU) 2017/745 on Medical Devices [2017] OJ L 117.
Regulation (EU) 2017/746 on in Vitro Diagnostic Medical Devices [2017] OJ L 117.
Regulation (EU) 2022/868 on European Data Governance [2022] OJ L 152.
Regulation (EU) 2024/1689 of the European Parliament and of the Council of 13 June 2024 laying down harmonised rules on artificial intelligence (Artificial Intelligence Act).
Regulation (EU) 2025/327 on the European Health Data Space and Amending Directive 2011/24/EU [2025] OJ L 327.
UN General Assembly, Convention on the Rights of Persons with Disabilities, 13 December 2006, A/RES/61/106.

2 Liability for Damages, AI, and Machine Learning for Digital Pathology as New Challenges for Biobanks

Dorota Krekora-Zając

1 Introduction

Algorithms, machine learning, and artificial intelligence (AI) have been employed in diagnostic processes for many years. More specifically, AI is used in digital pathology (Shafi and Parwani 2023), for scientific purposes or biobanking (Frascarelli et al. 2023). Digital pathology, for instance, opens new opportunities for biobanking and creates new legal challenges. Thanks to tissue and cell imaging, biobank collections can be made more widely available by digitizing biological samples such as pediatric samples, brain cells, or rare anomalies. The creation of digital tissue images in biobanks generates large datasets that are necessary to train AI and create diagnostic tools for further advancing digital pathology. Moreover, the adoption of European Health Data Space (EHDS) and the Artificial Intelligence Act (AI Act) will expectantly facilitate the free use of such datasets for scientific purposes and artificial intelligence exercises. Therefore, we are currently facing a challenge related, on the one hand, to easier access to large datasets originating from human biological samples and generated using artificial intelligence, and, on the other hand, to a change in the approach to the secondary processing of personal data in the EU. The aim of this chapter is to analyze whether there are principles of liability for damages in existing European legal acts and to indicate the most important legal issues in this respect.

2 What Is Digital Pathology?

M. G. Hanna and L. Pantanowitz (2019) define digital pathology as the practice of pathology using digital imaging; usage has employed static images, live streaming of images, and most recently whole slide imaging (WSI). WSI permits digitization of entire glass slides, thereby eliminating the need for pathology practice to be locally tethered to a conventional microscope. Practically, it consists of two processes. The first is based on creating digital images from microscopic preparations using a scanner. In the second one, pathologists use specialized software (dedicated browsers) for viewing and analyzing digital images (Leong and Leong 2004; Pantanowitz et al. 2011). Digital pathology can be broadly defined as creating, viewing, managing, sharing, analyzing, and interpreting digital images of microscopic

DOI: 10.4324/9781032634487-4

preparations. These applications include, but are not limited to, education, scientific research, archiving, telepathology/teleconsultations (Prochorec-Sobieszek and Szumera-Cieśkiewicz 2017), and routine diagnostics (Prochorec-Sobieszek 2016). The global Digital Pathology Market is projected to grow from USD 927.2 Mn in 2021 to 1748.1 Mn by 2030, at a CAGR of 7.3% in forecast period, 2022–2030.[1] Therefore, the rapidly developing digital pathology market will also result in a greater need to test diagnostic and scientific tools for digital pathology applications on data that is provided by biobanks.

3 Shortage of Pathologists and Use of AI Tools

In many countries, it is pointed out that there is a lack of pathologists.[2] Students rarely choose this specialization (D'Abbronzo et al. 2023). In Poland, for instance, reports indicate that the number of pathologists is so small that the waiting time for test results is prolonged, even delaying patients' opportunities to start treatment.[3] This has many further consequences. The shortage of physicians and improper financing of medical procedures led to the fact that the report of the Polish Supreme Audit Office shows that the Polish Institute of Oncology reported that 90% of the histopathological tests performed in other entities, both public and private (with which patients came for the first time to visit the clinics), included a description of the pathology findings and diagnosis, which could not constitute the basis for making a decision on the patient's treatment method (Supreme Audit Office 2020). The shortage of pathologists will also affect biobanks and scientific research. Incorrect results of pathological findings, for instance, that are transferred to biobanks, will consequently result in the creation of false data for scientific purposes. It is likely that exercising AI algorithms on such data may multiply erroneous data. Therefore, to enable the creation of reliable data also for biobanks, the use of digital pathology tools may burden laboratory diagnosticians and require data quality checks.

It is believed (Patrzyk and Woźniacka 2022) that thanks to the introduction of digital pathology equipment, from microscopes to imaging scanners and allowing digitalization of whole slides, the path has been opened to automated pathological diagnostics, supporting the physician with AI tools. Image analysis and machine learning systems have provided pathologists with decision support tools (Hanna and Ardon 2023). This will enable patients to access diagnoses more easily and help scientists obtain reliable data from biobanks.

4 Use Data for Digital Pathology

Digital images are used for many clinical and nonclinical applications. Clinical uses include making primary pathology diagnoses directly from digital images, telepathology, archiving, and image analysis. Nonclinical uses include quality testing, research, and education. There are several barriers to the widespread adoption of digital pathology including technology limitations, regulatory obstacles, cost of equipment, difficulties with implementation, and pathologist technophobia. Digital pathology is nevertheless deemed to become increasingly vital for the future practice of pathology (Hanna and Pantanowitz 2019).

AI and machine learning medical devices have the potential to be transformative in healthcare delivery, however, this change will only be realized if accompanied by effective governance that ensures patient safety, personal data protection, and public trust (Gilbert et al. 2023; Aboy et al. 2024; Ho and Caals 2024).

One example of the use of convolutional neural networks in histopathological diagnosis is the automatic determination of lung cancer subtypes. Used for this purpose, 1,635 paintings of the so-called virtual microscopy (whole slide imaging, WSI) were retrieved in a dataset from the Cancer Genome Atlas (TCGA). In this way, an algorithm was created that can assess with high probability whether the lesion is benign, an adenomatous cancer, or a squamous cell carcinoma (Chang et al. 2019).

Analysis of the usefulness of AI in histopathological diagnosis indicates a huge role of AI tools also in assessing effectively the advancement of the cancer process. Classification systems (staging), such as TNM (tumor, node, metastasis) in oncology, are based on data collected earlier. These include information on the degree of cancer development and invasion, lymph node involvement, and the presence of metastases, among other factors. Patient classification into specific categories allows for appropriate selection of treatment and helps in assessing the prognosis and prediction of survival. Staging – because it is based on qualitative descriptions that are often difficult to quantify – was, for some time, not perceived as suitable for automation. Automated determination of the size of the primary tumor and surrounding lymphatic nodes and distant metastases requires a combination of different methods. It turns out that deep learning, through the ability to learn and analyze multiple data simultaneously, is useful in solving also multifaceted classification problems. Consequently, AI has been used early on to detect whether the lymph nodes were involved or if distant metastases were present. To obtain an accurate assessment of the disease progression, it was trained on histopathological images (Patrzyk and Woźniacka 2022).

In addition, elaboration of AI-based computer models to diagnose and predict outcomes may facilitate clinical decision-making in accordance with the idea of personalized medicine (Robertson et al. 2018). For example, convolutional networks trained on appropriate resources (images and other data from the TCGA database) can distinguish glioblastoma multiforme from low glioma with high accuracy, including determining the level of differentiation between grades 2 and 3 (Chang et al. 2019). These techniques are used not only for oncological diseases. Xu et al. (2016) reported multimedia deep learning that yielded 87.83% sensitivity for cervical dysplasia.

AI algorithms in medicine are not commonly used as stand-alone software systems for making decisions. Rather, researchers tend to believe that the cooperation of artificial and human intelligence improves physicians' actions and reduces the risk of making mistakes (Sulaieva et al. 2024). Therefore, the use of AI tools may accelerate the work of pathologists as well as help to address the shortage of the same.

Another important application of digital pathology is biobanking. Thanks to digital pathology techniques, it is possible to conduct research not only on the biological sample itself but also on its images and the data derived from it. This enables wider access to data, especially when the biobanked material is irreplaceable. Therefore, the necessity to digitize biobank collections has been emphasized for years (Bonizzi et al. 2021). An example for such a unique and digitized biobank is the brain biobank, which comprises over 6,000 brain fragments, tumor biopsies, and other tissue samples taken from deceased donors or biological material obtained during surgeries, located at the Institute of Psychiatry and Neurology in Warsaw.[4]

5 Liability for Damages From EHDS

The basis for creating and using AI tools in the field of digital pathology are large medical datasets. Harmonization of the rules for the processing of personal data under the General Data Protection Regulation (GDPR) does not solve the problems in many countries regarding access to medical data for scientific purposes (Tzortzatou et al. 2021). In particular, in countries where no specific regulations regarding biobanking have been adopted (e.g., Poland), work has begun on codes of conduct for the processing of personal data, which aim to facilitate access to data on the one hand, while guaranteeing patients' privacy rights on the other hand (Krekora-Zając et al. 2021). Among other aspects, the European Union has long noticed problems with access to European medical databases as a key issue, especially for secondary use and biobanking. Pilot studies were, for instance, carried out under the HealthData@EU Pilot project and showed that national databases are not yet prepared to use data for secondary use of data in scientific research.[5] One of the significant challenges in this regard was the lack of a uniform legal framework. Difficulties in using medical data for scientific purposes became one of the reasons for starting work on the EHDS, which was formally adopted by the Council of the European Union on 25 January 2025, and entered into force in March the same year. Digital pathology may also become an important way of providing data within the EDHS. This is best demonstrated by the HSBOOSTER EU program, thanks to which it is possible to create synergy between various projects within the future EHDS framework (Gyrard et al. 2024), including the Bigpicture IMI project (H2020 2021–2027). Bigpicture collects and shares quality-controlled pathology images, and metadata, and relies on AI algorithms.[6]

It is suggested that establishing uniform standards for generating data in digital pathology can enhance the utilization of data within EHDS and become the gold standard in the field of precision medicine in the European Union (Nikolski et al. 2024). Akyüz et al. rightly point out that

> at a time when secure health data environments, such as European Health Data Space, are in the making, we argue that such meta-infrastructures can benefit both from the experience and evolution of biobanking, but also the current state of affairs in AI in medicine, regarding good governance, the

> social aspects and practices, as well as critical thinking about data practices, which can contribute to trustworthiness of such meta-infrastructures.
>
> (2024, 1)

The EHDS will enable almost unlimited access to various types of medical data based on Article 53 of the EHDS. It may seem that this regulation is revolutionary compared to the GDPR, which, however, significantly restricted the processing of medical data, for example, by requiring patient consent. The key difference between the GDPR and the EHDS is that data processed for scientific purposes under the EHDS will not allow the identification of the data subject. European regulations grant Member States considerable discretion to introduce more restrictive legal provisions, particularly regarding biobanks. In this respect, Member States have significant decision-making power over the availability of such data (Article 51(4) of the EHDS).

For this reason, the rules for sharing the data may be different, for example, in some countries, they are based on consent, in others, they are not (Marelli et al. 2023). Data quality will also be a significant challenge in this regard. In most EU countries, medical databases are not yet standardized, and data are saved in various formats (Dadaczynski et al. 2021; Vrdelja et al. 2021; Lievevrouw et al. 2024). In practice, this may generate many errors when using such files for scientific or diagnostic purposes, especially when using AI tools (Ranjbar and Ravn 2023).

The EHDS is a framework aimed, among other things, at enabling the sharing of data for training algorithms and artificial intelligence tools, also from biobanks. It does, however, not regulate the quality of such tools or data. The rules of such liability will therefore be determined by private law regulations of the Member States.

6 Liability for Damages in European AI Legislation

The issue of responsibility for AI is subject of the AI Act. It should, however, be noted at the outset that this is not a comprehensive regulation. It gives Member States great flexibility to create liability rules. On 12 July 2024, the AI Act (EU) 2024/1689 was published in the official journal of the European Commission and will be fully applicable 36 months after its entry into force – in August 2027. Actors must comply with the first rules by February 2025 (Article 113). This milestone is particularly significant for the healthcare sector, where the use of AI products for diagnosis, treatment, and patient care is rapidly increasing (van Kolfschooten and Oirschot 2024). The Regulation is a rather general act that leaves many possibilities for modification to the Member States (Gilbert 2024). For this reason, it is not fully clear what detailed rules of liability for damages related to the use of AI in digital pathology would be based on it.

The Act divides the application of AI into four risk categories. First, applications and systems that create an unacceptable risk, such as social credit scoring, are banned. Second is high-risk applications. Third is general-purpose AI as Large Language Models and fourth low-risk AI system, for example, health apps (see also chapter, "A Proposal for a Combined Risk Management Approach to AI Applications in Health Research and Biobanking" by Ilaria Colussi).

Pursuant to recital (50) of the preamble and Article 6, AI systems contained in medical devices are treated as high-risk systems. Risk is the combination of the probability of an occurrence of harm and the severity of that harm (Article 2(2)). The AI Act defines an AI system as a machine-based system that is designed to operate with varying levels of autonomy and that may exhibit adaptiveness after deployment, and that, for explicit or implicit objectives, infers, from the input it receives, how to generate outputs such as predictions, content, recommendations, or decisions that can influence physical or virtual environments (Article 3(1)). Therefore, the AI system does not acquire legal personality and is a tool that only enables humans to make better diagnoses (Buch et al. 2018). High-risk AI diagnostic tools must comply with certain safety and quality requirements, and operators have specific obligations regarding their use, including the EU Regulation on medical devices.

To minimize potential harm, the AI Act sets out rules for data management of high-risk AI systems which make use of techniques involving the training of AI models with data (Article 10). High-risk AI systems shall be designed and developed in such a way, including with appropriate human–machine interface tools, that they can be effectively overseen by natural persons during the period in which they are in use. Human oversight shall aim to prevent or minimize the risks to health, safety, or fundamental rights that may emerge when a high-risk AI system is used in accordance with its intended purpose or under conditions of reasonably foreseeable misuse (Article 14).

According to Article 2(8) AI Act, the regulation does not apply to any research, testing, or development activity regarding AI systems or AI models prior to their being placed on the market or put into service. Such activities shall be conducted in accordance with applicable EU law. Some point out that this exemption potentially leaves room to escape the regime for AI used in medical research and clinical trials, which may be a risk to patients (Colonna 2023).

Testing in real-world conditions shall not be covered by that exclusion. In relation to real-world testing, the AI Act establishes a requirement to obtain participant consent (Article 61). This consent should be freely given and informed. Subjects of testing prior to their participation in such testing have to be duly informed with concise, clear, relevant, and understandable information regarding not only the nature and conditions of the test but also the rights, and the guarantees regarding their participation, in particular their right to refuse to participate in, and the right to withdraw from, testing in real-world conditions at any time without any resulting detriment and without having to provide any justification.

According to Article 85 AI Act without prejudice to other administrative or judicial remedies, any natural or legal person having grounds to consider that there has been an infringement of the provisions of the Regulation may submit complaints to the relevant market surveillance authority. AI Act also provides for the possibility of imposing financial penalties (Article 99) and administrative fines on EU institutions, bodies, offices, and agencies (Article 100).

It is important to notice that the AI Act, in a very limited way, regulates patients' rights. There is no patient's right to object to a health professional using an AI system to diagnose or employ AI tools for other healthcare purposes. Moreover, Article 86 introduces the right to explanation of individual decision-making, which entails "the right to obtain from the deployer clear and meaningful explanations of the role of the AI system in the decision-making procedure and the main elements of the decision". This right also does not apply to AI used in medical devices and is only triggered if the AI decision has a significantly adverse impact on the health, safety, or fundamental rights of the patient. In practice, this means that in the context of healthcare, this right only exists in relation to AI decisions concerning access to essential benefits and (healthcare) services, life and health insurance, emergency calls or the dispatch of emergency first response services, and emergency healthcare triaging. It is important to note that the right to informed consent, as protected in all EU Member States, does entitle the patient to information about the medical treatment in a manner that is sufficient to make an informed decision about whether to proceed (van Kolfschooten and Oirschot 2024).

To sum up, it should be noted that AI Act does not provide for any specific rules of liability for tools that can be used in digital pathology. The AI Act, of course, imposes a number of obligations on producers to ensure patient safety, but it does not create a new liability regime.

On 19 September 2024, the European Parliamentary Research Service (EPRS) published a proposal for a directive on adapting noncontractual civil liability rules to AI – complementary impact assessment (Hacker 2024). The proposal highlights challenges relating to the Artificial Intelligence Liability Directive (AILD) and the revised Product Liability Directive (PLD). In October 2020, the European Parliament adopted its own resolution on AI liability (the European Parliament resolution of 2020 on AI liability) formally asking the European Commission to adopt a framework on AI liability based on a two-track approach: truly strict liability for high-risk AI systems, and a presumption of negligence for all other AI systems. Work on the directive is still at too early a stage to be able to conduct a legal analysis. Therefore, it will be omitted in further analyses.

It is clear from the earlier considerations that there are no pan-European standards regarding liability for erroneous data used for machine learning or the use of artificial intelligence in the field of digital pathology. Despite the lack of binding legal solutions, digital pathology creates new opportunities for both the development of medical services and scientific research.

There are three problems in terms of liability. The first concerns the possibility of limiting liability for incorrect data toward those conducting scientific research or preparing data for such research (nonmedical use). The second is the permissible limitation of liability toward the patient.

7 Problem With Legal Liability for Faulty Data in Research

The main problem regarding liability for incorrect data used for nonmedical purposes is the issue of various regulations regarding their origin. Sometimes tissue images may be treated as elements of medical documentation, sometimes as personal data, and in other cases as data not subject to these regimes. This is important for liability for incorrect data. The responsibility of a biobank or other entity that provides self-prepared tissue images will be different from that when they come from medical records kept, for example, by a hospital. Since this division refers to the interface between European and national law, this issue will be presented in the example of Polish law.

The legal basis for collecting data for medical purposes by doctors is Article 41(2) of Act on the Profession of Physicians and Dentists,[7] according to which the doctor is obliged to keep individual medical records of the patient. Keeping documentation in this area constitutes the processing of special categories of personal data under the GDPR. The data contained in the documentation are protected under the patient's right to privacy and may be transferred to other entities only in exceptional cases. Pursuant to Article 26(3)(2) of the Act on Patient Rights' and Patient Rights' Ombudsman,[8] such documentation may be transferred to public authorities, the National Health Fund, medical professions self-government bodies, and national and provincial consultants, to the extent necessary to perform by these entities their tasks, in particular control and supervision. The Act therefore explicitly allows for the transfer of medical records for health policy purposes without the need to anonymize the data derived from them.

The rules for data processing for scientific purposes are regulated completely differently. Pursuant to Article 26(4) of the Act on Patient Rights' and Patient Rights' Ombudsman, medical documentation may also be made available to a university or research institute for use for scientific purposes, without revealing the name and other data enabling identification of the person to whom the documentation relates. The doctrine indicates that a patient's objection to making his records available for scientific purposes is not binding, although it should be "considered" by the entity providing health services. Therefore, in accordance with the Act on Patient Rights' and Patient Rights' Ombudsman, the data in question may be transferred for scientific purposes, only as anonymized data, without the patient's additional consent. It seems that this regulation is consistent with the results of research on the need to respect the patient's consent to use his data for scientific purposes (Polanin and Terzian 2019).

Therefore, if the digital pathology test was performed for medical purposes, this image will be subject to medical documentation regulations. In practice, this will mean that the entity providing such documentation for scientific research purposes will not be able to modify its content. If only an image of the sample is transferred without a histopathological sample, the biobank will not be able to detect a possible error, for example, in terms of diagnosis. In such a case, it will be difficult to claim liability for faulty data. The recipient of data from the biobank will not be able to demonstrate either that the biobank is responsible for incorrect data (the biobank has no influence on them) or lack of due diligence in the performance of the contract.

In practice, the provisions of the Data Transfer Agreement (DTA) will play a key role in this respect. The agreement may limit or exclude liability for incorrect data in this respect or extend it, similarly to liability under warranty in a sales contract (Pursuant to Article 556 of the Civil Code), the seller is liable to the buyer if the sold item has any defect (warranty). However, it is indicated that basing liability on DTA creates significant problems in the case of data transfer to countries with different legislation (Swales et al. 2024; Swales et al. 2023). This problem is best illustrated by the discussion regarding data transfer from the EU to the USA (Lalova-Spinks et al. 2024).

The biobank's liability may be slightly different when it digitalizes samples on its own. The case law of Polish courts (Judgment of the Polish Supreme Court 2011) indicates that the paraffin block is part of the patient's medical documentation. However, it seems that such legal qualification cannot be extended to every biological sample in the biobank. In accordance with the principle of accuracy arising from the GDPR, the biobank is obliged to take reasonable steps to ensure the rectification or deletion of incorrect or outdated information (Nordberg 2021). However, there is still much controversy over whether a biobank can always be considered a data controller (Parziale and Staunton 2022).

Classically, it can be indicated that the biobank will then be contractually responsible for the proper quality of data. However, the question arises what responsibility to assign to a biobank that generates data using AI. There seems to be an adoption of accountability principles at European Union level in this case. Digital pathology and AI have caused the processed data to become detached from the carrier, which is a biological sample. And it will often be impossible to prove the sequence of events and entities responsible for the damage caused in the form of incorrect data.

The situation becomes even more complicated when the images created as part of digital pathology were anonymized and then transferred for scientific purposes. There is no legal regulation in this respect. GDPR then does not apply. The principles of liability will therefore result only from data transfer agreements. In the situation of machine learning and the creation of large databases for its purposes on the basis of various contracts, the real possibility of pursuing liability for faulty data will be very difficult. Additionally, there will be a high risk of reidentifying patients (Simon et al. 2019), which the data provider will not have much influence on (in the case of connecting databases and using AI). This problem is visible in many research projects in which it is necessary to appoint special committees to examine the possibility of reidentifying patients (Kondylakis et al. 2024).

8 Civil Liability in Poland for Incorrect Digital Pathology Results Obtained Using AI

Currently, there is no directly binding regulation of European law that would create an independent basis for compensation for damage related to the use of AI and machine learning in digital pathology. This means that the principles of liability are regulated by national laws in the liability regime or for medical damages or tort liability, including liability for a medical device (based on Regulation (EU) 2017/745) and a dangerous product (Directive 2001/95/EC).

AI Act clearly states that AI tools do not have their own legal entity, so the responsibility for incorrect results will fall on the physician, pathologist, biobanker, data provider, and entity introducing software or placing a medical device or product on the market. The implementation of the EHDS will bring significant changes for biobanks and medical entities in their approach to address the liability for false medical data. This will be especially relevant when a biobank operates as an organizational unit of a hospital, as a doctor's liability to the patient will extend to false data, including those used for scientific research and biobanking purposes. When creating datasets for EHDS, digital pathology tools will be essential. Consequently, pathologists working in hospitals and biobanks will face substantial challenges in ensuring the accuracy of their research results and upholding their professional responsibility.

Medicolegal liability is an issue of concern for all physicians, not only pathologists (Peterson 2019). Historically, a significant number of pathology errors precipitating malpractice suits have been those made in the analytic phase (i.e., false-positive and false-negative diagnoses), which conceivably could be both intercepted, and caused by AI tool contributions (Kornstein and Byrne 2007).

Research of D. B. Troxel (2006) shows that the false diagnoses that resulted in the most medical malpractice claims against pathologists most often concerned melanoma, breast cancer, and ovarian and prostate tumors. This may also be an important issue in relation to the risks associated with running a biobank (Samuel and Lucassen 2023).

Several concepts have been considered in the Polish law regarding liability for damages related to the use of AI. The first one assumes that the physician is responsible for any tool he uses in connection with the medical procedure he is conducting. In this respect, the obligation to carefully inspect the technical condition of the equipment used was pointed out each time (Judgment of the Polish Supreme Court 1983). However, it seems that the obligation to inspect a device such as a surgical knife does not correspond to the obligation to supervise AI. Similarly, a pathologist working in a biobank will not be able to be responsible for the quality of the equipment.

It was also pointed out that the need to take care of medical equipment rests with the healthcare entity (Shipley 1967). The manufacturer, importer, or distributor of the equipment was indicated as the third responsible entity. In this respect, obligations arising from the Regulation on medical devices and liability for a dangerous product are indicated. Traditionally, it can be indicated that the physician will be liable for defective treatment under the general principles adopted in a given country. *De lege lata*, there is no basis in this respect to treat the results obtained using AI in digital pathology as a reason to limit the physician's responsibility toward the patient, nor to impose limits on the biobank's liability.

However, the question arises whether the physician's liability will be modified if she informs the patient about the use of AI in his diagnosis. AI Act requires information in a very narrow scope. It is not known whether member states will attempt to impose such an obligation on physicians or pathologists working in a biobank.

Undoubtedly, discussions on this issue are ongoing, the best example of which is the attempt to regulate this issue in the Polish Code of Medical Ethics.

In Poland, the National Act on Artificial Intelligence, which is to implement the European AI Act, is at the stage of public consultations, and its draft has not yet been published. However, without waiting for the adoption of the act, the medical community began a discussion on the development of artificial intelligence tools and its impact on the understanding of a physician's duties. On 1 January 2025, the amended Article 12 of the Code of Medical Ethics (short KEL) enters into force, according to which the physician will have to ask the patient for consent to each use of AI in diagnostics.

According to this regulation, a physician may use it in diagnostic, therapeutic, or preventive procedures from AI algorithms when the following conditions are met: 1) Informing the patient that AI will be used when making a diagnosis or in the therapeutic process. 2) Obtaining the patient's informed consent to the use of AI in the diagnostic or therapeutic process. 3) The use of AI algorithms that are approved for medical use and have appropriate certificates. 4) The final diagnostic and therapeutic decision is always made by a physician. The justification for the change[9] stated that it was an expression of respect for the patient's autonomy in terms of respecting the patient's right to conscious participation in decisions regarding their health, including informing the patient about the benefits and risks of the proposed diagnostic and treatment activities, as well as about the possibilities of using other medical procedures.

The following principles related to such consent are indicated: Firstly, the patient's informed consent to the use of AI may be expressed orally or by the patient's behavior which clearly indicates his will – for example, lying on the examination couch, and continuing the conversation with the physician (in accordance with Article 16(1)(8) KEL). Article 12 of the KEL does not introduce a requirement to obtain consent in writing. Secondly, consent to the use of AI does not have to be separate from informed consent to the provision of a health service – if patients have been adequately informed about the use of AI in the diagnostic or therapeutic process, their consent to the provision of a health service will also include consent to the use of AI to provide this service. Thirdly, in accordance with the standards for obtaining informed consent, the information provided to patients about the use of AI in the diagnostic or therapeutic process should be formulated in a way that is understandable to them. In practice, this most often means that the physician should provide the patient with general information about the role of AI in the process and the importance of the information provided by AI for the entire process.

The use of AI does not release a person practicing a medical profession from the necessity to make a final professional assessment of the medical situation while observing the rules of caution within the framework of due diligence and taking into account the indications of current medical knowledge and the principles of professional ethics. One is obliged not only to know these conduct directives but also to improve their knowledge. Therefore, a representative of the medical profession must not blindly rely on automatic recommendations, even if their high accuracy has been confirmed by scientific research or his professional experience

(Kaczan 2022). Similarly, the use of AI in biobanking will not automatically release the biobank from liability.

Faulty programming, insufficient data, and the inability to fully predict how the device will behave in specific conditions may lead to personal injury in the form of bodily injury or disturbance to the patient's health. Errors may appear already at the diagnostic stage (e.g., incorrect pixel spacing on the scan may lead to the disease not being detected or a disease being diagnosed that does not actually exist) (Bączyk-Rozwadowska 2022).

Assigning liability to the manufacturer for damage caused by a device equipped with an AI system does not exclude the liability of other entities, both users and third parties (e.g., hackers), on other legal grounds (Kim 2018). The level of protection provided to patients is undoubtedly higher when they can submit compensation claims to several entities, each of which – until full satisfaction – remains obliged to full compensation (Bączyk-Rozwadowska 2022). Likewise, it will not free the biobank from liability.

These entities should be held jointly and severally liable with the manufacturer if they can be blamed for, among others, failure to perform required software updates or failure to take other actions aimed at maintaining the safe operation of the device (Liivak and Lahe 2018). However, this issue is particularly difficult when AI is used in digital pathology. The specificity of such diagnostics is that the doctor who obtains the result will not always have the opportunity to verify its correctness. Despite this *de lega lata*, it will not be possible to limit his liability toward the patient. Such a limitation of liability could only result from generally applicable law and not ethical codes. Similarly, in the absence of a clear legal regulation of biobanking in Poland, it will not be possible to limit the liability of a pathologist working in a biobank in this respect either.

Undoubtedly, a new challenge for medical law will be to define the principles of liability for disclosing patient data. According to Article 34, EHDS health data can be processed for secondary use of data also for training, testing, and evaluating of algorithms, including in medical devices, AI systems, and digital health applications, contributing to the public health or social security, or ensuring high levels of quality and safety of healthcare, of medicinal products, or of medical devices. This means that the data entered by the physician in the medical documentation will also be made available for the earlier purposes (Horgan et al. 2022). When a physician enters incorrect data, it may lead to the creation of faulty new tools, which may lead to harm not only to his patient but also to all other patients who will be diagnosed with the new tool in the future. The lack of detailed liability rules in European law may then lead to excessive liability (unlimited only to your own patient). According to Lucas J de Frutos and HT Haugo (2024), the burden of this responsibility should not fall on healthcare professionals, but specific limitations on liability are necessary in this regard.

Another problem related to reidentification is the loss of medical confidentiality. Especially when the obligation to disclose medical records does not result from the patient's consent but directly from a legal regulation, for example, EHDS. In this respect, in addition to ethical regulations (Staunton et al. 2024), it will be necessary to adopt European legal regulations defining the principles of this responsibility (Rodríguez-Mejías et al. 2024).

9 Conclusion

To sum up, it should be pointed out that the increasing use of AI in digital pathology creates new challenges for the law, in particular in the field of liability for damages. With regard to the use of AI in creating data for scientific research purposes, currently, all regulations regarding liability for incorrect data result directly from the DTA. However, contractual mechanisms will not be effective in the event of combining databases or obtaining data on another legal basis. In this regard, sharing data from digital pathology in EHDS will be a great challenge. One of the goals of the EHDS is to enable conducting scientific research, including research on AI. Nevertheless, the regulation does not provide any liability rules for incorrect data. This makes it difficult to determine what liability the providers of such data will have to bear.

The national legislators will also face great challenges in adopting a legal framework for AI tools in digital pathology for patients. It is indicated that legal regulations in this area should enable the development of patients' trust in new diagnostic methods (Borondy Kitts 2023). Nonetheless, the implementation of commercial healthcare AI faces serious privacy challenges (Murdoch 2021; Emam et al. 2009). *De lege lata*, AI is treated as a tool in the physician's work, therefore, it will not modify his liability for medical errors. Undoubtedly, however, the principle of physician responsibility will be the responsibility for disclosing data. EHDS in this respect will require a new look at the concept of patient data privacy and doctor's professional secrecy. *De lege lata*, the principles of biobanker liability should also be understood in a similar way. The mere use of AI tools in biobank facilities does not exempt us from responsibility for transmitting false scientific data.

Acknowledgments

This project has received funding from the National Science Center Poland "Agreements for biobanking of human biological samples for scientific purposes" (2016/23/D/HS5/00411).

Notes

1 Digital Pathology Market Size, Share & Trends Analysis, By Product (Device, Storage System), By End-use (Hospitals, Diagnostic Labs), By Application (Drug Discovery & Development, Academic Research), Region and Forecast Period 2022–2030 (Updated Version Available), https://marketresearchcommunity.com/digital-pathology-market/. (04.04.2025).

2 https://time.com/6234125/forensic-pathologist-shortage/ (08.04.2025).

3 www.rynekzdrowia.pl/Serwis-Onkologia/W-Polsce-jeden-patomorfolog-przypada-na-80-tys-mieszkancow-Prof-Marszalek-cos-zaczyna-sie-zmieniac,244211,1013.html (08.04.2025).

4 Digital Brain Biobank: https://digitalbrain.ipin.edu.pl/en/about-us/ (08.04.2025).
5 HealthData@EU Pilot identifies common elements for health data access and data use within the legal frameworks of the participating nodes, https://ehds2pilot.eu/wp-content/uploads/2023/07/Website-publication_2023_WP7-landscape-analysis_rev-1-1.pdf (08.04.2025).
6 https://bigpicture.eu/what-bigpicture-about (08.04.2025).
7 Ustawa z dnia 5 grudnia 1996 r. o zawodach lekarza i lekarza dentysty Dz.U. z 2024 r., poz. 1287.
8 Ustawa z dnia 6 listopada 2008 r. o prawach pacjenta i rzeczniku praw pacjenta Dz.U. z 2024 r. poz. 581.
9 Code of Medical Ethics justification for change: https://nil.org.pl/aktualnosci/8582-projekt-komentarza-art-12-kel-czekamy-na-propozycje-zmian. (21.03.2025).

Reference List

Aboy, Mateo, Timo Minssen, and Effy Vayena. 2024. "Navigating the EU AI Act: Implications for Regulated Digital Medical Products." *NPJ Digital Medicine* 7 (1): 237. https://doi.org/10.1038/s41746-024-01232-3.

Akyüz, Kaya, Mónica Cano Abadía, Melanie Goisauf, and Michaela Th. Mayrhofer. 2024. "Unlocking the Potential of Big Data and AI in Medicine: Insights from Biobanking." *Frontiers in Medicine* 11. https://doi.org/10.3389/fmed.2024.1336588.

Bączyk-Rozwadowska, Kinga. 2022. "Odpowiedzialność Cywilna Za Szkody wyrządzone W związku Z Zastosowaniem Sztucznej Inteligencji W Medycynie." *Przegląd Prawa Medycznego* 3 (3–4). https://doi.org/10.70537/z7xnk378.

Bonizzi, G., L. Zattoni, and N. Fusco. 2021. "Biobanking in the Digital Pathology Era." *Oncology Research* 29 (4): 229–33. https://doi.org/10.32604/or.2022.024892.

Borondy Kitts, Andrea. 2023. "Patient Perspectives on Artificial Intelligence in Radiology." *Journal of the American College of Radiology* 20 (9): 863–7. https://doi.org/10.1016/j.jacr.2023.05.017.

Buch, Varun H., Irfan Ahmed, and Mahiben Maruthappu. 2018. "Artificial Intelligence in Medicine: Current Trends and Future Possibilities." *British Journal of General Practice* 68 (668): 143–4. https://doi.org/10.3399/bjgp18X695213.

Chang, H. Y., C. K. Jung, J. I. Woo, et al. 2019. "Artificial Intelligence in Pathology." *Journal of Pathology and Translational Medicine* 53 (1): 1–12. https://doi.org/10.4132/jptm.2018.12.16.

Colonna, Liane. 2023. "The AI Act's Research Exemption: A Mechanism for Regulatory Arbitrage?" In *The Yearbook of Socio-Economic Constitutions: Law and the Governance of Artificial Intelligence*, edited by Andreas Moberg and Eduardo Gill-Pedro. Cham: Springer.

D'Abbronzo, G., S. Lucà, E. Carraturo, R. Franco, and A. Ronchi. 2023. "Shortage of Pathologists in Italy: Survey of Students and Residents." *Pathologica* 115 (3): 172–80. https://doi.org/10.32074/1591-951x-852.

Dadaczynski, Kevin, Orkan Okan, Melanie Messer, et al. 2021. "Digital Health Literacy and Web-Based Information-Seeking Behaviors of University Students in Germany During the COVID-19 Pandemic: Cross-sectional Survey Study." *Journal of Medical Internet Research* 23 (1): e24097. https://doi.org/10.2196/24097.

de Frutos Lucas, J., and H. T. Haugo. 2024. "Moving Forward with the European Health Data Space: The Need to Restore Trust in European Health Systems." *The Lancet Regional Health – Europe* 40: 100906. https://doi.org/10.1016/j.lanepe.2024.100906.

Frascarelli, C., G. Bonizzi, C. R. Musico, E. Mane, C. Cassi, E. Guerini Rocco, A. Farina, A. Scarpa, R. Lawlor, L. Reggiani Bonetti, S. Caramaschi, A. Eccher, S. Marletta, and N. Fusco. 2023. "Revolutionizing Cancer Research: The Impact of Artificial Intelligence in Digital Biobanking." *Journal of Personalized Medicine* 13 (9):1390. https://10.3390/jpm13091390.

Emam, K. E., F. K. Dankar, R. Vaillancourt, T. Roffey, and M. Lysyk. 2009. "Evaluating the Risk of Re-identification of Patients from Hospital Prescription Records." *Canadian Journal of Hospital Pharmacy* 62 (4): 307–19. https://doi.org/10.4212/cjhp.v62i4.812.
Gilbert, Stephen. 2024. "The EU Passes the AI Act and Its Implications for Digital Medicine Are Unclear." *NPJ Digital Medicine* 7 (1): 135. https://doi.org/10.1038/s41746-024-01116-6.
Gilbert, Stephen, Stuart Anderson, Martin Daumer, Phoebe Li, Tom Melvin, and Robin Williams. 2023. "Learning from Experience and Finding the Right Balance in the Governance of Artificial Intelligence and Digital Health Technologies." *Journal of Medical Internet Research* 25: e43682. https://doi.org/10.2196/43682.
Gyrard, A., P. Gribbon, R. Hussein, et al. 2024. "Synergies Among Health Data Projects with Cancer Use Cases Based on Health Standards." *Stud Health Technol Inform* 316: 1292–6. https://doi.org/10.3233/shti240649.
Hacker, Philipp. 2024. *Proposal for a Directive on Adapting Non-Contractual Civil Liability Rules to Artificial Intelligence. Complementary Impact Assessment*. European Parliamentary Research Service (EPRS). www.europarl.europa.eu/RegData/etudes/STUD/2024/762861/EPRS_STU(2024)762861_EN.pdf (08.04.2025).
Hanna, Matthew G., and O. Ardon. 2023. "Digital Pathology Systems Enabling Quality Patient Care." *Genes Chromosomes Cancer* 62 (11): 685–97. https://doi.org/10.1002/gcc.23192.
Hanna, Matthew G., and Liron Pantanowitz. 2019. "Digital Pathology." In *Encyclopedia of Biomedical Engineering*, edited by Roger Narayan, 524–32. Oxford: Elsevier.
Ho, Calvin Wai-Loon, and Karel Caals. 2024. "How the EU AI Act Seeks to Establish an Epistemic Environment of Trust." *Asian Bioethics Review* 16 (3): 345–72. https://doi.org/10.1007/s41649-024-00304-6.
Horgan, D., M. Hajduch, M. Vrana, et al. 2022. "European Health Data Space-An Opportunity Now to Grasp the Future of Data-Driven Healthcare." *Healthcare (Basel)* 10 (9). https://doi.org/10.3390/healthcare10091629.
Kaczan, Damian. 2022. "Odpowiedzialność odszkodowawcza za funkcjonowanie sztucznej inteligencji w medycynie." *Annales Universitatis Paedagogicae Cracoviensis | Studia de Cultura* 14 (2): 145–60. https://doi.org/10.24917/20837275.14.2.10.
Kim, Sunghyo. 2018. "Crashed Software: Assessing Product Liability for Software Defects in Automated Vehicles." *Duke Law & Technology Review* 16: 300–17.
Kondylakis, Haridimos, Rocio Catalan, Sara Martinez Alabart, et al. 2024. "Documenting the De-Identification Process of Clinical and Imaging Data for AI for Health Imaging Projects." *Insights into Imaging* 15 (1): 130. https://doi.org/10.1186/s13244-024-01711-x.
Kornstein, M. J., and S. P. Byrne. 2007. "The Medicolegal Aspect of Error in Pathology: A Search of Jury Verdicts and Settlements." *Archives of Pathology & Laboratory Medicine* 131 (4): 615–18. https://doi.org/10.5858/2007-131-615-tmaoei.
Krekora-Zając, Dorota, Błażej Marciniak, and Jakub Pawlikowski. 2021. "Recommendations for Creating Codes of Conduct for Processing Personal Data in Biobanking Based on the GDPR Art.40." *Frontiers in Genetics* 12. https://doi.org/10.3389/fgene.2021.711614.
Lalova-Spinks, Teodora, Peggy Valcke, John P. A. Ioannidis, and Isabelle Huys. 2024. "EU-US Data Transfers: An Enduring Challenge for Health Research Collaborations." *NPJ Digital Medicine* 7 (1): 215. https://doi.org/10.1038/s41746-024-01205-6.
Leong, F. J., and A. S. Leong. 2004. "Digital Photography in Anatomical Pathology." *Journal of Postgraduate Medicine* 50 (1): 62–9.
Lievevrouw, E., L. Marelli, and I. Van Hoyweghen. 2024. "Weaving EU Digital Health Policy into National Healthcare Practices. The Making of a Reimbursement Standard for Digital Health Technologies in Belgium." *Social Science & Medicine* 346: 116620. https://doi.org/10.1016/j.socscimed.2024.116620.
Liivak, Taivo, and Janno Lahe. 2018. "Delictual Liability for Damage Caused by Fully Autonomous Vehicles: The Estonian Perspective." *Masaryk University Journal of Law and Technology* 12 (1): 49–73.

Marelli, L., M. Stevens, T. Sharon, et al. 2023. "The European Health Data Space: Too Big to Succeed?" *Health Policy* 135: 104861. https://doi.org/10.1016/j.healthpol.2023.104861.

Murdoch, Blake. 2021. "Privacy and Artificial Intelligence: Challenges for Protecting Health Information in a New Era." *BMC Medical Ethics* 22 (1): 122. https://doi.org/10.1186/s12910-021-00687-3.

Nikolski, Macha, Eivind Hovig, Fatima Al-Shahrour, et al. 2024. "Roadmap for a European Cancer Data Management and Precision Medicine Infrastructure." *Nature Cancer* 5 (3): 367–72. https://doi.org/10.1038/s43018-023-00717-6.

Nordberg, Ana. 2021. "Biobank and Biomedical Research: Responsibilities of Controllers and Processors Under the EU General Data Protection Regulation." In *GDPR and Biobanking: Individual Rights, Public Interest and Research Regulation Across Europe*, edited by Santa Slokenberga, Olga Tzortzatou and Jane Reichel, 61–89. Cham: Springer International Publishing.

Pantanowitz, Liran, Paul N. Valenstein, Andrew J. Evans, et al. 2011. "Review of the Current State of Whole Slide Imaging in Pathology." *Journal of Pathology Informatics* 2: 36. https://doi.org/10.4103/2153-3539.83746.

Parziale, Andrea, and Ciara Staunton. 2022. "Research Biobanks and External Researchers Under the European General Data Protection Regulation: Between Controller-Processor Relationship and Joint Controllership." In *Integrity of Scientific Research: Fraud, Misconduct and Fake News in the Academic, Medical and Social Environment*, edited by Joel Faintuch and Salomão Faintuch, 93–100. Cham: Springer International Publishing.

Patrzyk, S., and A. Woźniacka. 2022. *Sztuczna inteligencja w medycynie*. Łódź.

Peterson, John E., Jr. 2019. "Survey of Medical Malpractice Cases and Settlements with Pathology or Laboratory Standard of Care Issues." *Seminars in Diagnostic Pathology* 36 (5): 366–71. https://doi.org/10.1053/j.semdp.2019.06.006.

Polanin, Joshua R., and Mary Terzian. 2019. "A Data-Sharing Agreement Helps to Increase Researchers' Willingness to Share Primary Data: Results from a Randomized Controlled Trial." *Journal of Clinical Epidemiology* 106: 60–9. https://doi.org/10.1016/j.jclinepi.2018.10.006.

Prochorec-Sobieszek, Monika. 2016. "Perspektywy rozwoju cyfrowej patomorfologii." *Biuletyn Polskiego Towarzystwa Onkologicznego Nowotwory* 1 (1): 1–9.

Prochorec-Sobieszek, Monika, and Anna Szumera-Ciećkiewicz. 2017. "Current State of Knowledge About Digital Pathology." *Hematology in Clinical Practice* 8 (1): 1–11. https://doi.org/10.5603/Hem.2017.0001.

Ranjbar, A., and J. Ravn. 2023. "Data Quality in Healthcare for the Purpose of Artificial Intelligence: A Case Study on ECG Digitalization." *Studies in Health Technology and Informatics* 305: 471–4. https://doi.org/10.3233/SHTI230534.

Robertson, S., H. Azizpour, K. Smith, and J. Hartman. 2018. "Digital Image Analysis in Breast Pathology-From Image Processing Techniques to Artificial Intelligence." *Translational Research* 194: 19–35. https://doi.org/10.1016/j.trsl.2017.10.010.

Rodríguez-Mejías, Silvia, Sara Degli-Esposti, Sara González-García, and Carlos Luis Parra-Calderón. 2024. "Toward the European Health Data Space: The IMPaCT-Data Secure Infrastructure for EHR-Based Precision Medicine Research." *Journal of Biomedical Informatics* 156: 104670. https://doi.org/10.1016/j.jbi.2024.104670.

Samuel, G., and A. Lucassen. 2023. "Access to Biobanks: Responsibilities Within a Research Ecosystem." *Biopreserv Biobank* 21 (3): 275–81. https://doi.org/10.1089/bio.2021.0172.

Shafi, S., and A. V. Parwani. 2023. "Artificial Intelligence in Diagnostic Pathology." *Diagnostic Pathology* 18: 109. https://doi.org/10.1186/s13000-023-01375-z.

Shipley, W. E. 1967. "Hospital's Liability to Patient for Injury Sustained from Defective Equipment Furnished by Hospital for Use in Diagnosis or Treatment of Patient." *14 ALR3d* 1254.

Simon, G. E., S. M. Shortreed, R. Y. Coley, et al. 2019. "Assessing and Minimizing Re-identification Risk in Research Data Derived from Health Care Records." *EGEMS (Wash DC)* 7 (1): 6. https://doi.org/10.5334/egems.270.

Staunton, Ciara, Mahsa Shabani, Deborah Mascalzoni, Signe Mežinska, and Santa Slokenberga. 2024. "Ethical and Social Reflections on the Proposed European Health Data Space." *European Journal of Human Genetics* 32 (5): 498–505. https://doi.org/10.1038/s41431-024-01543-9.

Sulaieva, Oksana, Oleksandr Dudin, Olena Koshyk, Mariia Panko, and Nazarii Kobyliak. 2024. "Digital Pathology Implementation in Cancer Diagnostics: Towards Informed Decision-Making." *Frontiers in Digital Health* 6. https://doi.org/10.3389/fdgth.2024.1358305.

Supreme Audit Office. 2020. "Organization, Availability and Quality Pathomorphological Diagnosis." KZD.430.008.2019, Nr ewid. 3/2020/P/19/062/KZD. www.nik.gov.pl/plik/id,22838,vp,25539.pdf (08.04.2025).

Swales, L., M. Botes, D. Donnelly, and D. Thaldar. 2023. "Towards a Data Transfer Agreement for the South African Research Community: The Empowerment Approach." *South African Journal of Bioethics and Law* 16 (1): 13–18. https://doi.org/10.7196/SAJBL.2023.v16i1.827.

Swales, Lee, Amy Gooden, and Donrich Thaldar. 2024. "The Anatomy of a Data Transfer Agreement for Health Research." *Frontiers in Pharmacology* 15. https://doi.org/10.3389/fphar.2024.1332700.

Troxel, David B. 2006. "Medicolegal Aspects of Error in Pathology." *Archives of Pathology & Laboratory Medicine* 130 (5): 617–19. https://doi.org/10.5858/2006-130-617-maoeip.

Tzortzatou, Olga, Santa Slokenberga, Jane Reichel, et al. 2021. "Biobanking Across Europe Post-GDPR: A Deliberately Fragmented Landscape." In *GDPR and Biobanking: Individual Rights, Public Interest and Research Regulation across Europe*, edited by Santa Slokenberga, Olga Tzortzatou and Jane Reichel, 397–419. Cham: Springer International Publishing.

van Kolfschooten, H., and J. van Oirschot. 2024. "The EU Artificial Intelligence Act (2024): Implications for Healthcare." *Health Policy* 149: 105152. https://doi.org/10.1016/j.healthpol.2024.105152.

Vrdelja, M., S. Vrbovšek, V. Klopčič, K. Dadaczynski, and O. Okan. 2021. "Facing the Growing COVID-19 Infodemic: Digital Health Literacy and Information-Seeking Behaviour of University Students in Slovenia." *International Journal of Environmental Research and Public Health* 18 (16). https://doi.org/10.3390/ijerph18168507.

Xu, Tao, Han Zhang, Xiaolei Huang, Shaoting Zhang, and Dimitris N. Metaxas. 2016. "Multimodal Deep Learning for Cervical Dysplasia Diagnosis." In *Medical Image Computing and Computer-Assisted Intervention – MICCAI 2016*. Cham: Springer.

Legal Sources

Code of Medical Ethics Justification for Change. https://nil.org.pl/aktualnosci/8582-projekt-komentarza-art-12-kel-czekamy-na-propozycje-zmian (21.03.2025).

Directive 2001/95/EC of 3 December 2001 on General Product Safety [2002] OJ L 11.

Judgment of the Polish Supreme Court of Feb. 9, 2011, V CSK 256/10. www.saos.org.pl/judgments/94195 (08.04.2025).

Judgment of the Polish Supreme Court of May 11, 1983, IV CR 118/8, OSNCPiUS 1983, nr 12, poz. 201.

Regulation (EU) 2017/745 on Medical Devices, Amending Directive 2001/83/EC, Regulation (EC) No 178/2002 and Regulation (EC) No 1223/2009 and Repealing Council Directives 90/385/EEC and 93/42/EEC [2017] OJ L117.

Regulation (EU) 2024/1689 Laying Down Harmonised Rules on Artificial Intelligence and Amending Regulations (EC) No 300/2008, (EU) No 167/2013, (EU) No 168/2013, (EU) 2018/858, (EU) 2018/1139 and (EU) 2019/2144 and Directives 2014/90/EU, (EU) 2016/797 and (EU) 2020/1828 (Artificial Intelligence Act) [2024] OJ L.

Ustawa z dnia 5 grudnia 1996 r. o zawodach lekarza i lekarza dentysty Dz.U. z 2024 r. poz. 1287.

Ustawa z dnia 6 listopada 2008 r. o prawach pacjenta i rzeczniku praw pacjenta Dz.U. z 2024 r. poz. 581.

Part 2

Between Risk and Trust

3 Unique by Design – the Need for Individualized Models of Trust in Biobanking AI

Anna Clareborn

1 Introduction

The continuous expansion of artificial intelligence (AI) into all facets of society is an undeniable reality, as general AI adoption and integration have increased drastically over the past couple of years (Gartner n.d.; Haenlein and Kaplan 2019; Mc Kendrick 2021). This implies exciting opportunities ahead, as we find ourselves at the beginning of what will doubtlessly be a major reimagining and restructuring of the current landscape of healthcare and biomedical sciences (Koski and Murphy 2021). Many argue that AI has the potential to improve healthcare and medical research to a significant degree, citing – among other benefits – precision, personalization, safety, speed, and the reduction of disparity (Bates 2021; Davenport 2019; Gruetzemacher and Whittlestone 2022; Green et al. 2013). As in the case of any significant technological advances in the medical and biomedical contexts, however, experts in the ethical, legal, and societal impact (ELSI) field find themselves playing catch-up. The struggle to keep up with these issues poses numerous regulatory challenges, and a vigorous discourse on the potential risks that accompany all this progress is a necessity to safeguard the protection of individual rights. How this discourse is advanced and maintained on various levels differs greatly around the world, however. In the EU, legislative developments are ongoing, as implementation is underway on the Artificial Intelligence Act (AI Act), the regulatory framework for AI that will affect nearly 450 million people in the 27 countries of the European Union (https://eur-lex.europa.eu/legal-content/EN/TXT/?uri=OJ%3AL_202401689).

Among the many potential hurdles ahead is the meaningful implementation of the AI Act in member countries where attitudes regarding the use of technology in general and AI specifically vary significantly. These attitudes closely tie in with issues of public trust, as well as the perceived trustworthiness of AI applications. Biobanking has a similarly complex connection to trust and trustworthiness, and the relevance of national and socioeconomic contexts cannot be refuted. But the EU is hardly homogeneous, and transnational implementations of regulatory measures present a significant challenge. There is wide-reaching consensus that the implications of this challenge must be addressed. There is less unanimity, however, regarding how to achieve an iteration of AI that takes these challenges into account and lives up to its suggested potential without significant fragmentation.

DOI: 10.4324/9781032634487-6

In this chapter, I start out by giving a background to the current status of the regulation and legislation of AI in Europe, focusing particularly on potential pitfalls regarding the practical implementation. I then discuss the balance between and interdependency of trust and trustworthiness, including the role of risk. A discussion regarding the relevance of trust and trustworthiness in AI is up next, with a particular focus on the pitfalls of actually implementing trustworthy AI. I then address the same topics, trust and trustworthiness, but in the context of biobanking, engaging with issues regarding risk factors and national disparities. AI applications in biobanking are considered next, offering a practical foundation for the topics of trust, trustworthiness, and risk. Finally, TAI, one of the suggested frameworks for the implementation of AI, is addressed, along with a concept that has been suggested as a potential resolution to some of TAI's inherent issues, namely co-production with a broad and inclusive base of AI stakeholders.

In terms of methodology used in the completion of the chapter, a selective literature review was conducted using systematic searches in PubMed, Scopus, as well as partially overlapping aggregate databases affiliated with Uppsala University Library and Oxford University Libraries. The main search terms applied relate to AI and biobanking ethics, biobanking AI applications, Trustworthy AI applications, and similar. Results were evaluated for relevancy, and an additional filter was used to assess how current the information was, particularly in the case of technological advances and applications. This yielded about 60% of sources, of which 99% consisted of peer-reviewed articles and other publications. About 40% of the sources used were previously known to the author from other contexts. In addition to previous types of materials, this category also consisted of official EU documents and similar.

2 Regulating and Legislating AI in Europe

In the European Union context, the High-Level Expert Group on AI (HLEG AI) produced its *Ethics Guidelines for Trustworthy Artificial Intelligence* (AI) on behalf of the European Commission in April 2019 (European Commission 2019). Health and well-being were singled out in the report as fields that present particularly relevant opportunities for the implementation of AI solutions. Biobanking was not mentioned specifically, but does seem to fall under what is referred to as areas where AI can "assist on a broader scale", with the ultimate goal of "more lives saved" (European Commission 2019). In February 2020, the follow-up publication, *White Paper on Artificial Intelligence: A European Approach to Excellence and Trust* was released, focusing particularly on the implementation of a new "ecosystem of trust" that should permeate all AI activities in the European Union. The paper also stressed the importance of adapting a risk-based approach to the implementation of regulatory measures regarding AI – healthcare applications of AI are used to exemplify various levels of risk, but the examples are general and do not address the extent or the complexities of the (bio)medical field.

The next step was legislation, and the European Commission announced its *Proposal for a Regulation on Artificial Intelligence* in April 2021. In accordance with the 2020 White Paper discussed earlier, AI applications were considered from the perspective of different levels of risk. The proposal also stressed the ecosystem of trust discussed earlier, and specifically emphasized the importance of ensuring that AI earns people's trust through the implementation of trustworthy AI.

Work on the proposal started off with a joint meeting of the lead committees at the end of January 2022. At that point, there were no clear indications regarding the extent to which healthcare AI in general, or biobanking AI more specifically, would be addressed in future legislative efforts. Only related research activities such as IVDMD/MD, or high-risk devices were considered directly (News European Parliament 2022). The White Paper strongly stressed the necessity for all member states to synchronize and streamline their AI initiatives and policies, which suggested potential clashes with current and upcoming localized AI guidelines and initiatives. Similar critical perspectives were offered from external sources as well, arguing that there was a significant risk that a wide variety of local AI regulations will be implemented in various EU member states in the vacuum of a consolidated EU regulation.

Different member states have different perceptions of AI and how it should be managed, as became clear in 2018, when the European Commission suggested a preliminary step toward AI consolidation that would ultimately result in all member states having implemented national AI strategies in 2019 – this, however, did not come to fruition (Larsson 2020). As we have seen in other contexts where local regulations are already extant, such as the GDPR, fragmentation can be significant and risks leading to equally significant challenges in terms of cooperation and synchronization further on (Borsci et al. 2022; Tzortzatou et al. 2021). The document acknowledged this fragmentation, emphasizing that this is something "Europe cannot afford to maintain" (White Paper Artificial Intelligence 2024). At the same time, however, no clear strategy for the consolidation and defragmentation of the AI landscape was available at that point.

The White Paper stressed that the HLEG AI Guidelines are a shared point of reference already in place, and one that will coexist with the AI Act (White Paper Artificial Intelligence 2024). It is reasonable to start out by examining some of those challenges described as particularly difficult to tackle in the context of AI cohesion, namely what has been referred to as "conceptual fuzziness" in the context of the Guidelines and the AI Act (Larsson 2020; Sako 2024).

In 2023, legislative efforts ramped up significantly, and on 12 July 2024, the AI Act was formally announced by being published in the Official Journal of the European Union. Finally, on 1 August 2024, it entered into force. However, the actual finality of this is debatable, as the requirements of the AI Act will only be applied and enforced gradually. On 2 February 2025, for instance, prohibitions on certain high-risk iterations of AI will begin to apply, and then several steps will follow. The final article, Article 6, will begin to apply on 2 August 2027, and deals with classification rules for high-risk AI systems (Artificial Intelligence Act 2024). It is not surprising that the bread in this legislative sandwich consists of regulations for high-risk systems – the past five years have been a global crash course in AI risk management, and EU legislators realize that time is needed to properly assess the implications of novel AI applications.

This process is an endeavor that will take time, not only because of the complexity of ever-changing AI regulatory needs but because the current legislation will also need to be consistently re-evaluated with the objective to ascertain whether or not it will suffice in its current form, or whether any additional legislative needs regarding high-risk AI can be covered in Article 6. It is up for debate what steps should be taken in the meantime in order to ensure a smooth transition from the current status quo to a shared framework in the field of biobanking AI, but it is reasonable to start by addressing the most relevant "fuzzy" cornerstones of AI acceptance – trust and trustworthiness.

3 Trust, Risk, and Trustworthiness

Trust is a highly complex concept that has been addressed from a multitude of perspectives in several academic disciplines, and no generally accepted definition or contextualization exists. Maintaining a broad perspective, however, it can be said that individuals carry beliefs and convictions about the world around them, and that these convictions dictate the extent to which unknown outcomes are interpreted as potentially positive or negative (Asan et al. 2020; Thiebes et al. 2021). Risk and risk-taking are central aspects of this, as trust can be described as the readiness of an individual (the trustor) to place trust in another individual, organization, or system (the trustee), thereby taking a risk in a situation where the outcome is uncertain. As soon as any degree of risk is involved, trust plays a significant role, implying that any exploration of one of the concepts must also address the other. As a matter of course, if trust has become damaged, placing trust in the same individual, organization, or system involves a higher degree of risk.

Trustworthiness, on the other hand, is an attribute connected to the trustee, in other words, the individual, organization, or system providing the product or service to the trustor. Trust in itself can be said to be entirely dependent on the degree to which the trustworthiness of the entity in question is subjectively experienced by the trustor. It has been argued that the formalization of trustworthiness into government structures muddles the ground for the understanding and operationalization of both trust and trustworthiness. More specifically, this implies that trustworthiness in its role as a governance structure carries an inherent risk of being reduced to mere signposting or tokenism. In other words, and in a worst-case scenario, slapping the label "trustworthy" on an organization or a system becomes an end goal in itself, and does not necessarily lead to an implementation of trustworthiness measures into the actual everyday practices of the entity in question (Samuel et al. 2022).

4 Trust and Trustworthiness in AI

There is a significant amount of research on trust in technology, but it is important to acknowledge that while the field of trust in AI is continuously expanding, there are still considerable gaps in the scholarship. A main reason that trust is such an integral aspect of the discourse on AI is the incontrovertible reality that no AI applications will be successful without securing the trust of the individuals who are intended to use them (Andtfolk et al. 2022; Bartneck et al. 2007; Calhoun et al. 2019; Lysø et al. 2024; Schepman et al. 2020). A loss of trust (either in the application itself, or, by proxy, in the institution behind it) directly correlates with a loss of users, and the awareness of this in the EU context is acute. In the context of biobanking, this would imply the loss of those samples that constitute the very basis for life-saving research. A general lack of trust in AI among European citizens is indeed one of the main issues cited in the White Paper as a reason why adoption of AI might turn out to be slow (White Paper Artificial Intelligence 2024).

Trust in technology or automation, such as AI applications, and in individuals or groups are generally characterized as different types of trust, and as such are subject to different theoretical frameworks (Cheng et al. 2022). Indeed, it has been argued that technology cannot be trusted at all; that it is something that at best can be relied on, and that the true trust relationship is with the various individuals, organizations, and other stakeholders involved in the development and maintenance of the technology in question (Simon and Rieder 2021).

In terms of AI, this implies an additional layer of complexity, since trustworthy AI, as will be discussed further below, may require various forms of human oversight. It is relevant to acknowledge that such human oversight is a key feature in the European Commission's White Paper, but that it is mainly perceived as relevant in relation to specifically high-risk AI applications, and as has already been clarified, biobanking is not explicitly considered a high-risk application at this stage in the legislative process (White Paper Artificial Intelligence 2024).

Looking at trust relationships in the case of AI, the trustor could be the user of the AI application, and the trustee could be the AI application itself. On this basic level, when activities are performed by the trustee (the AI application), risks pertaining to the trustor (the individual) might be realized (Castelfranchi and Falcone 2024). More complex interrelationships must also be considered, however, for instance, when tasks are delegated by the biobank, or by biobank researchers, to the AI application, which then interacts with sample donors or research participants on their behalf. Just as is the case with any system, the AI applications used by biobanks or researchers will hardly be developed by the biobanks in question. This implies an additional complication in the trust/trustworthiness relationship between the individual donating samples to the biobank and participating in the study. If the biobank utilizes an AI system to – for instance – inform patients about and/or collect consent, the responsibility for any realized risks is up for debate. This type of risk could consist of the meaning of consent being misunderstood by a patient,[1] or the wrong consent being registered in the system, which are both serious issues that could be harmful for trust.

These potential complexities aside, the most obvious trust/trustworthiness relationships in biobanking AI are reasonably between the biobank and the individual donating samples to the biobank, or between a researcher or research team and the individuals participating in a study. According to the trust/risk interdependency discussed earlier, an understanding of correlated risks is pivotal in order to evaluate any models or frameworks of trust. Kevin Anthony Hoff talks about "facilitating appropriate trust", implying there is a sweet spot where trust and risk are balanced to an ideal extent (Hoff and Bashir 2015, 407).

Navigation to that sweet spot, however, is dependent on understanding what is at stake. This is typically where transparency plays a role. Transparent iterations of AI have a high degree of explainability – simply put, the extent to which it is explainable how the AI works. But as Christian Lovis (2019, 4) puts it, "explainability might not be the best road to raise global trust in [aspects of AI], especially when the explanations themselves are hard to explain." In other words, *explainable* does not necessarily mean *understandable*. The black box nature of AI is an additional challenge to transparency and trust. Additionally, many have found that cultural context plays a large role in individuals' attitudes and perceptions regarding different iterations of AI (such as chatbots or robots), and that national differences can be highly significant (Haring et al. 2014). Additionally, if the cultural context is one where there is a low degree of trust in either technology in general or in the organization or entity behind the AI system in question, trustworthiness may be compromised in spite of good trustworthiness practices being in place.

5 Trust and Trustworthiness in Biobanking

To understand the role of trust and trustworthiness in biobanking, it is necessary to first examine some aspects of public perception in biobanking. The 2010 Eurobarometer and the inclusion of questions on public attitudes regarding biobanking and the participation in biomedical research was a watershed moment, considering the scarcity of such large-scale, comprehensive, pan-European data. A particularly significant takeaway from the survey results was the heterogeneity in terms of the perception of biobanking in different national contexts, and that nearly two-thirds of Europeans had not heard of biobanks at all. Only 2% discussed biobanking issues or retrieved information on biobanking with any frequency (European Commission 2010).

It is also relevant to consider new research initiatives during the 12 years since the Eurobarometer results were published. No efforts have been made on a similar scale, but several smaller national studies have added to insights about public perception. In Italy, the willingness to donate biobank samples was surveyed among family members of cognitively impaired patients, suggesting the need to differentiate between representatives of the public with varying degrees of insight into and investment in biobanking research (Porteri et al. 2014).

A German study with the goal of refreshing data from the 2010 Eurobarometer found that the majority of respondents have a positive attitude toward medical research in general. There were also positive attitudes toward biobanks, but even if respondents were willing to participate in biobank research and/or had a positive assessment of biobanking activities, it is important to note that out of these respondents, only around 30% had previously heard of biobanks (Bossert et al. 2018). Alas, not much notoriety or fame gained for biobanking since 2010 – at least not in Germany. Finally, an ambitious Latvian study was conducted in 2020, correlating the new results with the data from the Eurobarometer. Public awareness regarding biobanking had shrunk even more, but there was some increase in the willingness to participate in biobanking research. Much like in the older Italian study, different subcategories of the public could be discerned (Mezinska et al. 2020).

Other forms of evaluative efforts have also been attempted. A comprehensive 2019 literature review of 61 empirical studies regarding social attitudes toward biobanking clearly demonstrated that differences between countries are significant (Domaradzki and Pawlikowski 2019). However, clarifications are needed in terms of the reasons behind these differences, as well as fresher numbers. In an evaluation from 2005, the Scandinavian countries, for instance, had a matching level of *knowledge about* biobank research and *willingness to participate* in biobank research – around 90% would have willingly acted as donors (Kettis-Lindblad et al. 2006). In Greece, however, the corresponding number was 4% – but the reasons behind such significant discrepancies are unclear, as are correlations between Healthcare AI/Biobanking AI, and AI in general. A more recent, large-scale study of public perceptions of AI in Greece, for instance, suggested that nearly 50% of Greeks are skeptic toward AI and nearly 30% even feel fearful or anxious when they hear the term. Interestingly however, out of those who do feel positive about AI, 72% emphasize medicine as one of the sectors where AI will be able to do the most good (Newsroom 2022).

The intricacies of these results are typical for evaluations of willingness to participate in biobank research – results vary significantly between various subsections of survey participants, suggesting that interpretation can pose a challenge (Bossert et al. 2018). It should also be noted that it has been suggested that diversity in terms of public opinions about biobanking can be far more complex than what cultural and/or geographic contextualization can suggest (Hoeyer 2010). A more in-depth analysis of this is beyond the scope of this chapter, but it is relevant to reflect upon the potential of AI in this context. As will be addressed in the next section, AI provides a significant opportunity to tailor certain applications and services to the individual user, something that would be impossible to do through more traditional means, mainly due to lack of resources. This could imply that AI is a potential tool to negotiate such complex diversities.

Risk awareness and management play a central role in the context of trust in biobanking as well (Johnsson 2013). In the ethical discourse on participation in (bio)medical research, the perception of biobanking, or the participation in biobanking research as a low-risk endeavor is generally accepted and rarely challenged. Minimally risky research has been described as research that "will obviously include minimally invasive and minimally risky procedures such as participation in biobanks" (Harris 2005, 247). It has also been characterized as including "participation in registries and biobanks" (Mascalzoni et al. 2017), and as "research that imposes only diminutive risks, including research using stored samples and data" (Stjernschantz et al. 2014, 328). It is also true that risks associated with biobanking are rarely physical in nature, as biobanking is typically not very invasive. Time sensitivity is also rarely an issue, and immediacy-induced risk scenarios, as the one explored earlier, are not as likely in biobanking as in general healthcare AI.

Instead, breaches of privacy are generally defined as the main risk, at worst resulting in sensitive personal information about research participants ending up in the wrong hands (Stjernschantz Forsberg et al. 2012). This corresponds with perceptions of biobanking held by potential or actual participants in biobank research. Commenting on a large-scale literature review of public perceptions of biobanking, Jan Domaradzki concludes that "for the majority of respondents, participation in a biobank was a risky enterprise as they were afraid of the possibility of linking biological samples with donors' personal data" (Domaradzki and Pawlikowski 2019, 5). It is of course necessary to differentiate between patients and research subjects in biobanking research. Accordingly, a research participant who is not a patient is subjected to a certain level of risk through participation in biobank (and, to a lesser degree registry) research. For a patient with a potentially stigmatizing diagnosis, however, risks are more tangible; some particularly strong fears expressed by the public relate to their dealings with, for instance, employers and insurance companies (Domaradzki and Pawlikowski 2019). It is important to emphasize that these are not just hypothetical concerns. There are examples of such fears actually coming to fruition, resulting in significant consequences for individuals (D'Abramo 2015).

Among the general AI risks stressed particularly in the European Commission's White Paper, however, are citizens' fear of being powerless, and fear of unintended effects and misuse of AI applications. Additionally, threats to the safety and health of individuals (such as loss of life) and immaterial threats (such as loss of privacy, bias, and discrimination) are discussed (White Paper Artificial Intelligence 2024). When AI risks do come to fruition, consequences can be dire, as a reduction of trust undermines the entire system (Forcier et al. 2019). These overreaching themes affect healthcare AI as well, but there are also healthcare-specific risks that pose significant challenges that might not affect biobanking directly. Examples of this are immediacy-induced risks, or risks that arise in immediate and pressing care situations. When a person's health is at stake, they might not consider potential long-term impacts of AI systems being used to deliver the help they need to be a priority. Accordingly, studies have shown that individuals who have been diagnosed with an illness are more likely to embrace an AI solution, in accordance with a "whatever helps" outlook (Bærøe et al. 2020). In order to discuss possible solutions, however, it is necessary to look more closely at current and potential future applications of AI in biobanking.

6 AI Applications in Biobanking

It is important to stress that the discourse on healthcare AI in general, biomedical AI, and biobanking AI more specifically is quite muddled. As AI applications relating to the biomedical context gain more traction, so will certainly the differentiation between general healthcare and (bio)medical AI and biobanking AI. Additional issues can be found in terms of the overall proliferation of research.

Some claim that AI healthcare-oriented scholarship is heavily biased toward US perspectives on AI attitudes and that more geographically inclusive research is needed (Hagerty and Rubinov 2019). There is also a significant need for more comprehensive research on biobanking AI applications, as well as risk scenarios and consequences.

Various applications of AI in the (bio)medical field have been implemented since the 1960s. In spite of this long history, the field is still at an early stage, but picking up significant pace in accordance with the current wave of AI development. It is reasonable to assume that the expansion of biobanking in AI will continue at a similar pace as (bio)medicine and healthcare AI in general. In a survey of current trends in medical AI, the four following main categories of practical use were defined: 1) assessing the risk of disease onset and estimating treatment success, 2) managing or alleviating complications, 3) ongoing patient care, and 4) ongoing pathology and treatment efficacy research (Becker 2019). Although some concurrence with the earlier categories is evident, no equally comprehensive review of applications of AI specific to the field of biobanking exists, but some efforts have been made to summarize current trends. One such summary from 2020 cites examples such as analysis of biobanked images, genome sequencing, and cross-analyses of samples and data for specific diagnoses (Kozlakidis 2020).

It has been argued that biobanking, with its tight overlap with precision medicine research and big data models, is a "natural extension for AI-driven activities", and AI has been defined as a necessary field of inclusion in biobanking training and education (Kozlakidis 2020; Kinkorová 2021). One systematic study of AI use in the international biobanking context concludes that AI presents a unique opportunity to elevate biomedical sciences, particularly in the field of personalized medicine. Additional uses for AI could be as a strategic tool for biomedical research by analyzing trends in research needs and correlation to biobank sample inventory (Battineni et al. 2022). These all represent those iterations of AI that are not necessarily obvious or known to us – nonetheless, we are surrounded and affected by them to a large extent in our daily lives.

It is particularly relevant, however, to consider obvious AI applications, that is, instances of AI that are immediately discernible to the individual. In biobanking today, this mainly involves various aspects of patient communication. The use of AI in this context, such as through a chatbot interface, is a growing field, but one that faces many challenges due to the limitations of the AI bots (Bhirud et al. 2019). An example of where AI could be applied in a biobanking situation is as a replacement of or an alternative to what is traditionally patient/clinician or research subject/researcher interactions. This could be in situations such as giving or receiving information, like informing about consent. It could also apply to an extended interaction where patients/subjects provide consent after having received information. Consent issues in biobanking are particularly complex as regards patient communication, setting biobanking apart from more general communication in the healthcare context. The emphasis in healthcare AI might be on limited, Q&A-oriented chatbots or their more anthropomorphic counterparts, the so-called embodied conversational agents (ECAs), who catalogue symptoms or remind individuals to take their medication at specific time intervals (ter Stal et al. 2020). In biobanking, however, needs are likely to veer more toward the gathering of data directly from patients/data subjects, or ensuring that consent that is given is indeed informed – and informed in a manner that has been tailored to suit the consent provider's individual needs, taking both ethical and legal dimensions of consent into account (Rao 2008).

For participants in research studies, for instance, a significant amount of information must be processed by the patient/subject recipient of that information. Depending on the context, this might take place in an environment that is often unfamiliar and with numerous external stressors. Clinicians and researchers are frequently pressed for time and might struggle to find a balance between treating patients/subjects equally, while at the same time considering individual needs. Research shows that AI applications can prove successful in terms of such patient/subject-centeredness, in other words, taking the individual's needs into account and engaging on a personal level, while entirely avoiding the power imbalance inherent in traditional interpersonal communication between clinicians/researchers and patients/subjects (Green and Stadler 2013).

One 2019 study regarding a consent, chatbot mirrors similarly positive results. The bot "walks patients through components of the consent allowing them to opt to receive more or less detail on key topics (goals, benefits, risks, etc.)". The chatbot also records the decision of the patient (Schmidlen et al. 2019, 1168). Patients' responses to the AI bot were positive throughout, particularly emphasizing such benefits as being able to control the speed and content of the information provided. They also found it easier to remember information presented in the app-based consent interaction than in a comparable interaction with a human being. The only concerns regarding the AI consent bot cited as relevant by the participants in the study were those regarding access to or skills in terms of using the chatbot technology (Schmidlen et al. 2019). As discussed earlier, individual conditions vary greatly between different national and socioeconomic contexts, and AI could present an opportunity to negotiate these various differences and needs, rendering biobanking more understandable and accessible to a larger group of participants. This is further supported by research on consumers' responses to chatbots in other settings – trust is in fact directly affected by the AI bot's ability "to appreciate their perspective and meet their specific needs" (Cheng et al. 2022, 508). This potential to build trust needs to be harnessed through the application of measures that optimize AI. Several challenges regarding such measures must be addressed, however, in order to realize the desired "ecosystem of trust".

7 Trustworthy AI (TAI) Frameworks and AI Co-Production

One suggested path toward such an ecosystem is the implementation of Trustworthy Artificial Intelligence (TAI). The definition of TAI, however, is subject to continuous re-evaluation and reinterpretation. The degree of readiness in terms of actually operationalizing TAI is also subject to discussion. One article states that the implementation of TAI is a stage where "companies are still in the beginning of this process or have not even started yet" (Baker-Brunnbauer 2021, 1). This, however, does not mean that there are not attempts at getting the process started; there are numerous suggested iterations of frameworks and tool kits.

One of the more impactful among contemporary models, AI4People, was launched in 2018 and is based on five foundational principles, namely beneficence, non-maleficence, autonomy, justice, and explicability (Floridi et al. 2018). This can be compared to the more streamlined categories in the HLEG AI Guidelines, outlining TAI as necessarily lawful, ethical, and robust (European Commission 2019). TAI frameworks are frequently criticized for either attempting to do too much or settling with doing too little. In the case of the HLEG AI Guidelines, as Larsson points out, the guidelines – in spite of emphasizing that they are not intended to cover legal issues – deal with areas that have a strong overlap with legislative topics (potentially doing too much) (Larsson 2020). At the same time, the document has been criticized for not providing a suitably comprehensive model for trustworthy AI (potentially doing too little) (Borsci et al. 2022). In spite of what must be assumed to be an attempt at an accessible and user-friendly format, the Guidelines are therefore already spawning meta-frameworks developed to help negotiate the pitfalls of the practical implementation, such as the Trustworthy AI Implementation Framework (TAII) (Baker-Brunnbauer 2021).

In this context, it is necessary to return to the previous discussion on the risk of trustworthiness taking the shape of a purely tokenist, signposting, or symbolic concept. Any TAI framework must contain a strategy for implementation, but the assessment and adjustment of compliance are just as important. Steps have been taken in this area as well. Roberto V Zicari and others point out that numerous studies delineate the lack of compliance with and adherence to ethical guidelines for AI, and furthermore emphasize that the "contradiction between having both an abundance of high-level guidance and unethical applications of AI is partly reflected in the need for systematic auditing methods aimed at assessing or increasing the trustworthiness of AI applications" (Zicari et al. 2021, 5). How to address this issue practically, however, proves a challenge. One recommendation is that "ethical inconsistencies during and after the development of AI systems [can be prevented by] the implementation of AI ethics . . . with the support of a multidisciplinary meta perspective by the key stakeholders" (Baker-Brunnbauer 2021, 2). It is incontrovertible that one such group of key stakeholders is made up of patients (in the case of healthcare AI) and other representatives of the public. However, PPI, or patient and public involvement in the development of AI systems is by no means an established practice. It has been suggested that civil society representatives have not actually been invited as participants in the EU dialogue on AI regulation and legislation, even though attempts to remedy this have been made recently (Digital Strategy 2024).

The importance of trust and trustworthiness cannot be downplayed in this case. Patient and/or public involvement, particularly through high-level inclusivity strategies like co-production, has been proven to help trust, not least when trust has been damaged or is at a low level for another reason (Abelson et al. 2003). It is unclear why PPI or other forms of public involvement are not implemented in AI development to a larger extent. Among reasons often cited in the healthcare or (bio) medical context, this type of involvement or engagement often suffers from a lack of awareness of power imbalances, as well as inadedquate follow-up and evaluation measures (Andress et al. 2020; Domecq et al. 2014). This could certainly be true for AI as well, but it is unclear to what extent measures have been taken to attempt involvement in the first place.

Another issue can be connected to the definition of representatives of the public as relevant stakeholders in AI. A comprehensive study focusing on involvement and co-production has shown that representatives of the public are generally excluded as stakeholders, even in initiatives clearly targeted at and/or involving the public, unless they are very clearly defined as key stakeholders from the very start of a project or initiative (Razavi et al. 2020). From a broader perspective, this suggests that we need to clearly designate patients and other representatives of the public as key stakeholders in healthcare AI (Hashiguchi et al. 2022). Narrowing that down to include biobanking that corresponds to individuals who already have donated or who will potentially donate samples to biobanks. Inclusion in this context should at the very least include the implementation of measures that allow sample donors to be involved in the process of evaluating and suggesting improvements of those AI iterations they interact with directly, such as chatbots and other types of "obvious" AI.

It has been argued, however, that we could set the bar even higher. This would be achieved by implementing co-production as a key feature of TAI, in order to contribute to trust and trustworthiness in healthcare and (bio)medical AI. Fully participatory design through co-production could be a potential option. This would mean involving patients and/or other representatives of the public from start to finish even in nonobvious AI, for instance, through the development of AI algorithms. A suggested added bonus of this more immersive approach to building public trust in biobanking is that it could take the edge off the dystopic/ panacea dichotomy that sometimes plagues the discourse on AI, rendering public perception of AI more balanced (Hashiguchi et al. 2022). It is necessary to emphasize, however, that the lack of consensus in the field of patient and public involvement is not likely to be mitigated in the context of AI. The addition of challenges like explainability will more likely add to this discord. Additional inclusion strategies would also be necessary to ensure representativity and build trust, particularly in the case of marginalized or other sensitive groups (Erikainen et al. 2021; Zicari et al. 2021).

8 Discussion

We still know too little about the upcoming implementation of the AI Act to be able to embrace the suggested "ecosystem of trust" fully in the context of biobanking AI. But what does the landscape of trust in biobanking AI look like, and how does it correlate to trust in healthcare AI more generally? As discussed earlier, trust demands risk awareness to come to fruition – and for risk awareness to be achieved, transparency is required. As a general rule, optimism, if not excitement, characterizes public opinion about the potential of medical AI in Europe and beyond, but the chasm is vast between those perceiving AI as a panacea and those defining it as a harbinger of an impending apocalypse. Across the board, however, medicine is perceived as one of the most promising areas for the implementation of AI. It is doubtful that everyone with this conviction grasps the intricacies of medical AI in greater detail, but *believing* you understand is generally enough for a positive outlook.

In terms of biobanking, there are numerous promising venues of application as well. However, several long-running issues hamper the field. One problem is familiarity. With members of the European public unaware of the mere concept of biobanking, and making no particular effort to find out more, this constitutes a potential roadblock on the path toward familiarity. The implementation of a Trusted AI framework, such as they are conceived today, would matter very little to the public if this were the case. Another problem is negativity – as has been stressed earlier – since AI applications operate on trust. No matter how trustworthy a TAI framework is, if it is applied to a field that is not trusted by the public, the result will be lackluster at best. An individual who does not want to provide samples or data to a biobank will most likely remain reluctant even if the TAI score is high.

This also spotlights the difficulties in collectively interpreting and addressing the data we do have about public opinion regarding biobanking. It is incontrovertible that there are significant differences between European countries in terms of the perception of biobanks in general, and the participation in biobank research more specifically. However, it is less clear what reasons lie behind these discrepancies. Correlations have been suggested between participation and high general knowledge about biobanks or biobanking research. However, as I have argued earlier, we cannot know if merely informing the public would be a viable campaign for boosting biobanking positivity in countries where participation is not the popular option, and where attitudes toward AI are less positive. Even in cases where we do know more about how general societal attitudes inform the perception of biobanking, there is still more trial and error required before viable strategies for addressing the issue can be implemented.

The current discourse on risk and risk management in (bio)medical applications of AI is dynamic, and it is clear that risk management is one of the most highly prioritized issues in the current regulatory activities in the European sector. As has been stressed earlier, the correlation between trust and risk comprehension is pivotal for the trustor/trustee relationship in the adoption and implementation of AI applications. In order for trustworthiness to be realized, concepts like transparency, explainability, and understandability must be addressed. It is obvious, however, that efforts to do this are hampered by a lack of homogeneity in terms of how these concepts are understood and applied. Even considering individual needs in, for instance, different national contexts, shared core principles for trust and trustworthiness are pivotal.

To summarize, there is a significant tension between the lack of public trust in the field of biobanking in several contexts, and the foundation of trust any AI initiative needs to build upon. Additionally, local discrepancies regarding trust and public perception indicate a need for enough flexibility to accommodate these variations. Furthermore, there is a lack of a shared frame of reference regarding core concepts. The implementation process for the AI Act has only just begun, but the outlined core challenges in biobanking do not fall under any high-risk provisions at this stage. As such, they are unlikely to be directly affected by Article 6 considerations. As fragmentation is likely to be a challenge in the wake of any legislative or regulatory changes, it would be more relevant to address what can be consolidated and streamlined at this stage. Laying the groundwork for a shared framework that could pave the way for the implementation of AI in biobanking would be a significant step forward, and one that could potentially mitigate the effects of unwanted regulatory consequences.

But such frameworks, potentially in the guise of TAI, are not yet the carefully sharpened tools they need to be to realize the full potential of AI in biobanking, particularly not if they should live up to the aspirations of excellence of the ecosystem of trust promoted throughout the legislative process of the AI Act. If we consider the discussion on formalized trustworthiness as signposting or tokenism, ensuring that trustworthiness as a concept is not eroded by inadequate operationalization and everyday practice appears to be a necessary step. Any TAI models must contain built-in processes for compliance assessment and evaluation. An additional concept that aids in circumnavigating tokenism is transparency, which, as we have seen, brings with it challenges of its own in terms of AI.

9 Conclusion

Even with those issues addressed, additional challenges remain. An inherent trait of an ecosystem is that all components are necessary, and that imbalance is an unavoidable consequence if one or several such components are removed. In light of that, it is relevant to consider how little research is being done in terms of how much patients and other members of the public have been involved in the development of AI tools in healthcare and (bio)medicine. This is especially pertinent, considering how much points to co-design being a potentially very powerful tool to boost current TAI frameworks by addressing key concerns. Co-production is a path toward building nonexistent trust or rebuilding diminished trust, as well as a facilitator of transparency. Looking at the practical application of biobanking, the addition of engagement and involvement practices could allow donors to evaluate and suggest improvements to AI tools. This would also open the door for a greater level of adaptation and individualization of these tools – this can be considered an additional potential advantage, as it could help negotiate the issue of individual differences between various national and/or socioeconomic contexts, augmenting AI to suit the context to a much larger degree. More research is certainly needed regarding how PPI and/or other forms of public involvement could be implemented in future TAI frameworks to achieve the Commission's visionary ecosystem of trust.

Note

1 It is important to acknowledge that not all samples in a biobank come from patients – individuals with a specific diagnosis. There are other donor types as well, such as healthy research participants. To simplify, however, I will use the term patient throughout, as these donors are the main focus of my research studying biobanks in a clinical setting.

Reference List

Abelson, Julia, Pierre-Gerlier Forest, John Eyles, Patricia Smith, Elisabeth Martin, and Francois-Pierre Gauvin. 2003. "Deliberations about Deliberative Methods: Issues in the Design and Evaluation of Public Participation Processes." *Social Science & Medicine* (1982) 57 (2): 239–51. https://doi.org/10.1016/s0277-9536(02)00343-x.

Andress, Lauri, Tristen Hall, Sheila Davis, Judith Levine, Kimberly Cripps, and Dominique Guinn. 2020. "Addressing Power Dynamics in Community-Engaged Research Partnerships." *Journal of Patient-Reported Outcomes* 4 (1): 24. https://doi.org/10.1186/s41687-020-00191-z.

Andtfolk, Malin, Linda Nyholm, Hilde Eide, Auvo Rauhala, and Lisbeth Fagerström. 2022. "Attitudes Toward the Use of Humanoid Robots in Healthcare – A Cross-Sectional Study." *AI & Society* 37 (4): 1739–48. https://doi.org/10.1007/s00146-021-01271-4.

Artificial Intelligence Act. 2024. "Implementation Timeline | EU Artificial Intelligence Act." https://artificialintelligenceact.eu/implementation-timeline/ (18.10.2024).

Asan, Onur, Alparslan Emrah Bayrak, and Avishek Choudhury. 2020. "Artificial Intelligence and Human Trust in Healthcare: Focus on Clinicians." *Journal of Medical Internet Research* 22 (6): e15154. https://doi.org/10.2196/15154.

Bærøe, Kristine, Ainar Miyata-Sturm, and Edmund Henden. 2020. "How to Achieve Trustworthy Artificial Intelligence for Health." *Bulletin of the World Health Organization* 98 (4): 257–62. https://doi.org/10.2471/BLT.19.237289.

Baker-Brunnbauer, Josef. 2021. "TAII Framework for Trustworthy AI Systems." *SSRN Scholarly Paper*. Rochester, NY: Social Science Research Network. https://papers.ssrn.com/abstract=3914105.

Bartneck, Christoph, Tomohiro Suzuki, Takayuki Kanda, and Tatsuya Nomura. 2007. "The Influence of People's Culture and Prior Experiences with Aibo on Their Attitude Towards Robots." *AI & Society* 21 (1): 217–30. https://doi.org/10.1007/s00146-006-0052-7.

Bates, David W., David Levine, Ania Syrowatka, et al. 2021. "The Potential of Artificial Intelligence to Improve Patient Safety: A Scoping Review." *NPJ Digital Medicine* 4 (1): 1–8. https://doi.org/10.1038/s41746-021-00423-6.

Battineni, Gopi, Mohmmad Amran Hossain, Nalini Chintalapudi, and Francesco Amenta. 2022. "A Survey on the Role of Artificial Intelligence in Biobanking Studies: A Systematic Review." *Diagnostics* (Basel, Switzerland) 12 (5): 1179. https://doi.org/10.3390/diagnostics12051179.

Becker, Aliza. 2019. "Artificial Intelligence in Medicine: What Is It Doing for Us Today?" *Health Policy and Technology* 8 (2): 198–205. https://doi.org/10.1016/j.hlpt.2019.03.004.

Bhirud, Nivedita, Subhash Tatale, Sayali Randive, and Shubham Nahar. 2019. "A Literature Review on Chatbots in Healthcare Domain." *International Journal of Scientific & Technology Research* 8 (7).

Borsci, Simone, Ville Lehtola, Francesco Nex, et al. 2022. "Embedding Artificial Intelligence in Society: Looking beyond the EU AI Master Plan Using the Culture Cycle." *AI & Society* 38: 1–20. https://doi.org/10.1007/s00146-021-01383-x.

Bossert, Sabine, Hannes Kahrass, and Daniel Strech. 2018. "The Public's Awareness of and Attitude Toward Research Biobanks – A Regional German Survey." *Frontiers in Genetics* 9: 190. https://doi.org/10.3389/fgene.2018.00190.

Calhoun, Christopher, Philip Bobko, Jennie Gallimore, and Joseph Lyons. 2019. "Linking Precursors of Interpersonal Trust to Human-Automation Trust: An Expanded Typology and Exploratory Experiment." *Journal of Trust Research* 9: 1–19. https://doi.org/10.1080/21515581.2019.1579730.

Castelfranchi, Christiano, and Rino Falcone. 2024. "Trust Is Much More than Subjective Probability: Mental Components and Sources of Trust | IEEE Conference Publication | IEEE Xplore." https://ieeexplore.ieee.org/document/926815/similar#similar (18.10.2024).

Cheng, Xusen, Ying Bao, Alex Zarifis, Wankun Gong, and Jian Mou. 2022. "Exploring Consumers' Response to Text-Based Chatbots in e-Commerce: The Moderating Role of Task Complexity and Chatbot Disclosure." *Internet Research* 32 (2): 496–517. https://doi.org/10.1108/INTR-08-2020-0460.

D'Abramo, Flavio. 2015. "Biobank Research, Informed Consent and Society. Towards a New Alliance?" *Journal of Epidemiology and Community Health* 69 (11): 1125–8. https://doi.org/10.1136/jech-2014-205215.

Davenport, Thomas, and Ravi Kalakota. 2019. "The Potential for Artificial Intelligence in Healthcare." *Future Healthcare Journal* 6 (2): 94. https://doi.org/10.7861/futurehosp.6-2-94.

Digital Strategy. 2024. "Commission Launches Consultation on AI Act Prohibitions and AI System Definition | Shaping Europe's Digital Future." https://digital-strategy.ec.europa.eu/en/news/commission-launches-consultation-ai-act-prohibitions-and-ai-system-definition.

Domaradzki, Jan, and Jakub Pawlikowski. 2019. "Public Attitudes toward Biobanking of Human Biological Material for Research Purposes: A Literature Review." *International Journal of Environmental Research and Public Health* 16 (12): 2209. https://doi.org/10.3390/ijerph16122209.

Domecq, Juan Pablo, Gabriela Prutsky, Tarig Elraiyah, et al. 2014. "Patient Engagement in Research: A Systematic Review." *BMC Health Services Research* 14: 89. https://doi.org/10.1186/1472-6963-14-89.

Erikainen, Sonja, Phoebe Friesen, Leah Rand, et al. 2021. "Public Involvement in the Governance of Population-Level Biomedical Research: Unresolved Questions and Future Directions." *Journal of Medical Ethics* 47 (7): 522–5. https://doi.org/10.1136/medethics-2020-106530.

European Commission. 2010. Directorate-General for Research and Innovation. Stares, Sally, Allansdottir, Agnes, and Gaskell, Geroge, Europeans and Biotechnology in 2010 – Winds of Change? Publications Office.

European Commission. 2019. Directorate-General for Communications Networks, Content and Technology. Ethics Guidelines for Trustworthy AI. Publications Office.

European Commission. 2020. White Paper of the European Commission of 19 February 2020 on Artificial Intelligence – A European Approach to Excellence and Trust COM(2020) 65 final.

Floridi, Luciano, Josh Cowls, Monica Beltrametti, et al. 2018. "AI4People – An Ethical Framework for a Good AI Society: Opportunities, Risks, Principles, and Recommendations." *Minds and Machines* 28 (4): 689–707. https://doi.org/10.1007/s11023-018-9482-5.

Forcier, Mélanie Bourassa, Hortense Gallois, Siobhan Mullan, and Yann Joly. 2019. "Integrating Artificial Intelligence into Health Care Through Data Access: Can the GDPR Act as a Beacon for Policymakers?" *Journal of Law and the Biosciences* 6 (1): 317–35. https://doi.org/10.1093/jlb/lsz013.

Gartner. n.d. "Gartner Forecasts Worldwide Artificial Intelligence Software Market to Reach $62 Billion in 2022." www.gartner.com/en/newsroom/press-releases/2021-11-22-gartner-forecasts-worldwide-artificial-intelligence-software-market-to-reach-62-billion-in-2022 (18.10.2024).

Green, Nancy L., Sara Rubinelli, Donia Scott, and Adriaan Visser. 2013. "Health Communication Meets Artificial Intelligence." *Patient Education and Counseling* 92 (2): 139–41. https://doi.org/10.1016/j.pec.2013.06.013.

Green, Nancy L., and Brian Stadler. 2013. "Adding Coping-Related Strategies to Biomedical Argumentation in Computer-Generated Genetic Counseling Patient Letters." *Patient Education and Counseling* 92 (2): 149–52. https://doi.org/10.1016/j.pec.2013.05.001.

Gruetzemacher, Ross, and Jess Whittlestone. 2022. "The Transformative Potential of Artificial Intelligence." *Futures* 135: 102884. https://doi.org/10.1016/j.futures.2021.102884.

Haenlein, Michael, and Andreas Kaplan. 2019. "A Brief History of Artificial Intelligence: On the Past, Present, and Future of Artificial Intelligence." *California Management Review* 61: https://doi.org/10.1177/0008125619864925.

Hagerty, Alexa, and Igor Rubinov. 2019. "Global AI Ethics: A Review of the Social Impacts and Ethical Implications of Artificial Intelligence." *arXiv*. https://doi.org/10.48550/arXiv.1907.07892.

Haring, Kerstin, Céline Mougenot, Fuminori Ono, and Katsumi Watanabe. 2014. "Cultural Differences in Perception and Attitude towards Robots." *International Journal of Affective Engineering* 13: 149–57. https://doi.org/10.5057/ijae.13.149.

Harris, J. 2005. "Scientific Research Is a Moral Duty." *Journal of Medical Ethics* 31 (4): 242–8. https://doi.org/10.1136/jme.2005.011973.

Hashiguchi, Tiago Cravo Oliveira, Jillian Oderkirk, and Luke Slawomirski. 2022. "Fulfilling the Promise of Artificial Intelligence in the Health Sector: Let's Get Real." *Value in Health: The Journal of the International Society for Pharmacoeconomics and Outcomes Research* 25 (3): 368–73. https://doi.org/10.1016/j.jval.2021.11.1369.

Hoeyer, Klaus. 2010. "Donors Perceptions of Consent to and Feedback from Biobank Research: Time to Acknowledge Diversity?" *Public Health Genomics* 13 (6): 345–52. https://doi.org/10.1159/000262329.

Hoff, Kevin Anthony, and Masooda Bashir. 2015. "Trust in Automation: Integrating Empirical Evidence on Factors That Influence Trust." *Human Factors* 57 (3): 407–34. https://doi.org/10.1177/0018720814547570.

Johnsson, Linus. 2013. "Trust in Biobank Research: Meaning and Moral Significance." PhD diss., Uppsala University. https://urn.kb.se/resolve?urn=urn:nbn:se:uu:diva-192295.

Kettis-Lindblad, Åsa, Lena Ring, Eva Viberth, and Mats G. Hansson. 2006. "Genetic Research and Donation of Tissue Samples to Biobanks. What Do Potential Sample Donors in the Swedish General Public Think?" *European Journal of Public Health* 16 (4): 433–40. https://doi.org/10.1093/eurpub/cki198.

Kinkorová, Judita. 2021. "Education for Future Biobankers – The State-of-the-Art and Outlook." *The EPMA Journal* 12 (1): 15–25. https://doi.org/10.1007/s13167-021-00234-5.

Koski, Eileen, and Judy Murphy. 2021. "AI in Healthcare." *Studies in Health Technology and Informatics* 284: 295–9. https://doi.org/10.3233/SHTI210726.

Kozlakidis, Zisis. 2020. "Biobanks and Biobank-Based Artificial Intelligence (AI) Implementation Through an International Lens." *Lecture Notes in Computer Science*: 195–203. https://doi.org/10.1007/978-3-030-50402-1_12.

Larsson, Stefan. 2020. "On the Governance of Artificial Intelligence through Ethics Guidelines." *Asian Journal of Law and Society* 7 (3): 437–51. https://doi.org/10.1017/als.2020.19.

Lovis, Christian. 2019. "Unlocking the Power of Artificial Intelligence and Big Data in Medicine." *Journal of Medical Internet Research* 21 (11): e16607. https://doi.org/10.2196/16607.

Lysø, Emilie Hybertsen, Maria Bårdsen Hesjedal, John-Arne Skolbekken, and Marit Solbjør. 2024. "Men's Sociotechnical Imaginaries of Artificial Intelligence for Prostate Cancer Diagnostics – A Focus Group Study." *Social Science & Medicine* (1982) 347: 116771. https://doi.org/10.1016/j.socscimed.2024.116771.

Mascalzoni, Deborah, Carlo Petrini, Domenica Taruscio, and Sabina Gainotti. 2017. "The Role of Solidarity(-Ies) in Rare Diseases Research." *Advances in Experimental Medicine and Biology* 1031: 589–604. https://doi.org/10.1007/978-3-319-67144-4_31.

Mc Kendrick, Joe. 2021. "AI Adoption Skyrocketed Over the Last 18 Months." https://hbr.org/2021/09/ai-adoption-skyrocketed-over-the-last-18-months (18.10.2024).

Mezinska, S., J. Kaleja, I. Mileiko, D. Santare, V. Rovite, and L. Tzivian. 2020. "Public Awareness of and Attitudes towards Research Biobanks in Latvia." *BMC Medical Ethics* 21 (1): 65. https://doi.org/10.1186/s12910-020-00506-1.

News European Parliament. 2022. "Artificial Intelligence Act: Lead Committees to Launch Joint Work on 25 January | News | European Parliament." www.europarl.europa.eu/news/en/press-room/20220119IPR21312/artificial-intelligence-act-lead-committees-to-launch-joint-work-on-25-january.

Newsroom. 2022. "What Greeks Think about the Potential and Dangers of Artificial Intelligence." *Emeastartups*. https://emeastartups.com/what-greeks-think-about-the-potential-and-dangers-of-artificial-intelligence/9295 (23.02.2022).

Porteri, Corinna, Patrizio Pasqualetti, Elena Togni, and Michael Parker. 2014. "Public's Attitudes on Participation in a Biobank for Research: An Italian Survey." *BMC Medical Ethics* 15: 81. https://doi.org/10.1186/1472-6939-15-81.

Rao, K. H. Satyanarayana. 2008. "Informed Consent: An Ethical Obligation or Legal Compulsion?" *Journal of Cutaneous and Aesthetic Surgery* 1 (1): 33. https://doi.org/10.4103/0974-2077.41159.

Razavi, S. Donya, Lydia Kapiriri, Michael Wilson, and Julia Abelson. 2020. "Applying Priority-Setting Frameworks: A Review of Public and Vulnerable Populations' Participation in Health-System Priority Setting." *Health Policy* (Amsterdam, Netherlands) 124 (2): 133–42. https://doi.org/10.1016/j.healthpol.2019.12.005.

Sako, Natalija. 2024. "EU's AI Act Could Face Implementation Challenges." www.chalmers.se/en/current/news/cse-eu-s-ai-act-could-face-implementation-challenges/ (14.03.2024).

Samuel, Gabrielle, Reinder Broekstra, Felix Gille, and Anneke Lucassen. 2022. "Public Trust and Trustworthiness in Biobanking: The Need for More Reflexivity." *Biopreservation and Biobanking* 20 (3): 291–6. https://doi.org/10.1089/bio.2021.0109.

Schepman, Astrid, and Paul Rodway. 2020. "Initial Validation of the General Attitudes towards Artificial Intelligence Scale." *Computers in Human Behavior Reports* 1: 100014. https://doi.org/10.1016/j.chbr.2020.100014.

Schmidlen, Tara, Marci Schwartz, Kristy DiLoreto, H. Lester Kirchner, and Amy C. Sturm. 2019. "Patient Assessment of Chatbots for the Scalable Delivery of Genetic Counseling." *Journal of Genetic Counseling* 28 (6): 1166–77. https://doi.org/10.1002/jgc4.1169.

Simon, Judith, and Gernot Rieder. 2021. "Trusting the Corona-Warn-App? Contemplations on Trust and Trustworthiness at the Intersection of Technology, Politics and Public Debate." *European Journal of Communication* 36 (4): 334–48. https://doi.org/10.1177/02673231211028377.

Stjernschantz Forsberg, Joanna. 2012. "Biobank Research: Individual Rights and Public Benefit." PhD diss., Uppsala University. https://urn.kb.se/resolve?urn=urn:nbn:se:uu:diva-171898.

Stjernschantz Forsberg, Joanna, Mats G. Hansson, and Stefan Eriksson. 2014. "Why Participating in (Certain) Scientific Research Is a Moral Duty." *Journal of Medical Ethics* 40 (5): 325–8. https://doi.org/10.1136/medethics-2012-100859.

ter Stal, Silke, Lean Leonie Kramer, Monique Tabak, Harm op den Akker, and Hermie Hermens. 2020. "Design Features of Embodied Conversational Agents in eHealth: A Literature Review." *International Journal of Human-Computer Studies* 138: 102409. https://doi.org/10.1016/j.ijhcs.2020.102409.

Thiebes, Scott, Sebastian Lins, and Ali Sunyaev. 2021. "Trustworthy Artificial Intelligence." *Electronic Markets* 31 (2): 447–64. https://doi.org/10.1007/s12525-020-00441-4.

Tzortzatou, Olga, Santa Slokenberga, Jane Reichel, et al. 2021. "Biobanking Across Europe Post-GDPR: A Deliberately Fragmented Landscape." In *GDPR and Biobanking: Individual Rights, Public Interest and Research Regulation across Europe*, edited by Santa Slokenberga, Olga Tzortzatou and Jane Reichel, 397–419. Cham: Springer International Publishing.

Zicari, R. V., J. Brodersen, J. Brusseau, et al. 2021. "Z-Inspection®: A Process to Assess Trustworthy AI." *IEEE Transactions on Technology and Society* 2 (2): 83–97. https://doi.org/10.1109/TTS.2021.3066209.

Legal Sources

Proposal of the European Commission of 21 April 2021 for a Regulation of the European Parliament and of the Council Laying Down Harmonised Rules on Artificial Intelligence (AI Act) and Amending Certain Union Legislative Acts COM(2021) 206 final. https://eur-lex.europa.eu/legal-content/EN/TXT/?uri=OJ%3AL_202401689.

4 A Possible Risk Governance Approach for AI in Health Research and Biobanking

Ilaria Colussi

1 Introduction

Artificial Intelligence (AI) is one of the fastest emerging technologies, which is radically transforming several fields of human activity. The replication of human intellectual processes has, indeed, the potentiality to impact on many areas, including health research and biobanking, where AI could help improve human understanding of diseases and diagnosis, the comprehension of data collected in biobanks, the acquisition of knowledge from data (Battineni et al. 2022), or help detect the quality of samples and data, and process quickly large quantities of data (Hussain et al. 2002; Mesko 2017).

However, AI also gives rise to some challenges and risks, related to computing powers, safety, security, privacy, transparency, equality, possible threats to exercising fundamental rights, and ethics principles (Kin 2024; Müller 2014).

Finding an effective approach of risk governance to AI "that can mitigate these risks and cope with the resulting challenge of ensuring safe, responsible, and beneficial development and deployment of AI" (Cihon et al. 2020, 228) is crucial. Therefore, the focus of this chapter is to identify a workable risk governance approach for AI technologies in the field of health research and biobanking in practice.

In literature, different approaches to risk governance of new technologies can be identified, but our choice has been devoted to five approaches that we deem useful for allowing a comprehensive comparative analysis of all the key elements essential for evaluating AI risks. These approaches are based on the Precautionary Principle, the Cost-Benefit Analysis, the Responsible Innovation, the Proactionary Principle, and the Innovation Principle (Hemphill 2020). Each principle represents a distinct way to balance risk mitigation and innovation, and offers diverse perspectives on how AI risks should be managed: so, we selected these five principles to have the possibility to best compare all the relevant aspects that are, in our opinion, needed when conducting risk governance.

By dissecting various governance models, their strengths, and limitations, methodologically through a SWOT analysis (Strengths, Weaknesses, Opportunities, and Threats), it becomes possible to identify gaps, propose refinements, and develop an approach that is both comprehensive and adaptable to the rapid pace of technological advancement. The added value of such an inquiry lies in its ability to

DOI: 10.4324/9781032634487-7

inform policymakers, researchers, and industry stakeholders on the most effective strategies for AI risk mitigation. Additionally, by aligning theoretical governance models with practical regulatory mechanisms, such an examination ensures that AI applications, especially in critical domains like health and biobanking, operate within a framework that safeguards fundamental rights, fosters public trust, and promotes responsible innovation.

The analysis of how different governance approaches can apply to AI is particularly relevant given the evolving regulatory landscape, where legal frameworks, such as the EU's Artificial Intelligence Act (AI Act), are still shaping the governance of AI risks. Indeed, a regulation/legislation cannot ignore the choice of a risk governance approach guiding it. In particular, the EU is the first entity in the world having decided to regulate AI with a legislative act. The recently adopted AI Act (21 May 2024, Regulation 2024/1689) aims to establish a robust legal framework to foster the development of secure, trustworthy, and ethical AI, and to introduce a common regulatory and legal framework for AI encompassing almost all sectors and applying rules to all types of AI. Since the AI Act follows a risk-based approach ("the higher the risk, the stricter the rules") and establishes that the level of scrutiny and compliance requirements for AI systems will be proportional to the level of risk they pose, there is a need to look into the risk governance approach that the EU has opted for, and examine whether the EU approach coincides with the approach proposed by this chapter, or whether some elements are missing.

In light of the earlier, this chapter is divided into the following sections: first of all, some examples of AI applications in the fields of health research and biobanking, together with the risks that AI gives rise to, will be provided; then, possible approaches of risk governance will be analyzed; thirdly, a workable "combined" approach will be suggested and presented; fourthly, the focus will be devoted to the EU AI Regulation, considering which approach of risk governance it takes and whether it is in line with the current proposal, or whether some aspects are lacking; finally, the suggested governance approach will be applied to health research and biobanking, in order to see its concrete application in the field.

1.1 Benefits and Risks in Health Research and Biobanking

AI in health research and biobanking poses many challenges and risks. The possible applications of AI in the area of health research and biobanking are growing (WHO 2021). Some applications entail the design, development, and evaluation of algorithmic techniques used to acquire knowledge and learn from data (the so-called "machine learning") (Craig and Tucci 2024). The algorithms of machine learning techniques are used by Electronic Health Records (EHR) (Zewe 2021). Random Forest, for example, is a powerful ensemble learning method used to predict the likelihood of patient readmission within 30 days after discharge (Michailidis et al. 2022), or Support Vector Machines (SVMs) for disease classification (Gao et al. 2023), and Natural Language Processing (NLP) for extracting insights from notes (Hossain et al. 2023). Machine learning is also adopted in the context of biomedical research to improve human understanding of a disease (Hussain et al. 2002).

As regards biobanking (Grossman and Henderson 2023; Rychnovská 2021), AI systems could help digitalize, analyze, and process large volumes of data (by leveraging various technologies that automate data extraction, categorize unstructured information, identify patterns, streamline data processing, and generate actionable insights with speed and accuracy). Moreover, AI can define and help to quantify the quality of biosamples; for example, the AI system will be able to detect the integrity of DNA (Waring et al. 2020), it can develop a strategy for future biomedical research by studying the biobanks distribution and inventory status and research trends (thus relocating biosamples to maximize storage space utilization, by shifting biospecimen collection and storage from manual to automated processes) (Ezziane 2006), and it can also lead to virtual biobanks that provide a single point of access to a network of biospecimens (CloudLIMS 2022).

While AI shows many potential benefits, it also raises risks and challenges related to its potentiality to affect human rights, freedoms, dignity, and implementation of ethical and legal principles (Tableau 2024; Frascarelli et al. 2023). AI may increase data threats, thus affecting data safety and data security. It requires large amounts of data to operate, but this makes research institutions more vulnerable to data breaches. AI algorithms can be more susceptible to cyberattacks and threaten data integrity and privacy (The Economic Times 2023; Chukhnov and Ivanov 2021). Furthermore, data used for research are usually fragmented across many different systems, and the fragmentations may lead to errors and less safety and security measures, thus decreasing the effectiveness and operability of AI systems (Crigger and Khoury 2019, 188).

There is the perception that AI can cause physical harm, related to property damage, injury, or even death: indeed,

> autonomous robots, capable of moving and acting in the physical world, are the most obvious, but not the only type of AI that can cause physical harm. It is also possible that decision-support systems can lead to physical damage. An AI system supporting medical professionals may provide inaccurate diagnoses.
>
> (Crigger and Khoury 2019, 188)

Additionally, AI can cause nonphysical harm, such as financial loss or a privacy violation, or psychological harm, and harm to dignity. For example, AI algorithms can also lead to biased decisions depending on the data with which AI has been trained, and this may generate discrimination, violations of diversity, and inequality effects, especially affecting vulnerable populations. AI systems learn from the data with which they are trained, and if those data contain biases, then AI may replicate the same inequalities that exist in healthcare systems (e.g., Afro-American patients receiving less treatments than white patients, or gender discrimination) (Bajorek 2019).

Another risk is the one linked to transparency, including accessibility, explainability, and explicability of AI systems: AI leads to results which are not always explainable and the decision-making and reasoning process contributing to these results is opaque. It is the so-called "black box" area (Yu and Alì 2019; Scherer 2015), affecting medicine too.

As regards privacy and confidentiality (Price and Cohen 2019), AI is based on the collection and processing of large amounts of data. If AI developers gather data from electronic health systems or even from third parties (such as insurance) without proper legal, technical, security, and organizational measures, people can feel that their privacy has been violated.

2 Different Risk Governance Approaches for AI

Since AI generates not only benefits but also risks, it is relevant to reflect on what approach to adopt in order to balance the development and realization of AI potential for health, while minimizing the risks. In other words, it is urgent to find a proper governance which prevents from hampering progress and innovation, but at the same time protects human rights, freedoms, and implementation of ethical/legal principles. Thus, a reflection on risk governance of AI is needed (Taeihagh 2021; Latonero 2018).

In general terms, a "risk governance" model usually consists of three phases: a) risk assessment, b) risk management, and c) risk communication (Grieger et al. 2023). First of all, it is needed to identify risks, then to decide how to manage them, and thirdly how to communicate them to the stakeholders, meaning the public, whose perception of risk is important for trust and acceptability of AI.

In literature according to Hemphill (2020), there are several principles or approaches that have been shaped for the risk management phase. Our choice has been devoted to five approaches which could be particularly suitable for evaluating AI:

a) The Precautionary Principle
b) The Cost-Benefit Analysis
c) The Responsible Innovation
d) The Proactionary Principle (also known as "Permissionless Innovation")
e) The Innovation Principle

These principles have been chosen because they appear particularly relevant for doing a comparative analysis and for providing a comprehensive evaluation of AI governance from multiple perspectives. Additionally, these principles are widely used in the governance of emerging technologies, making them applicable, adaptable, and well-suited to AI. Other frameworks might not provide the necessary comparative analysis between precaution, safety, risk-taking, cost-effectiveness, and innovation incentives. Thus, these five principles were selected not only because they address the key concerns surrounding AI governance but also because they enable a meaningful comparison of different governance strategies, making them the most relevant for this analysis.

2.1 Precautionary Principle

The Precautionary Principle has been translated into several versions. It is common to distinguish between a "strong" version and a "weak" one. The first one affirms,

> When an activity raises threats of harm to the environment or human health, precautionary measures should be taken even if some cause-and-effect relationships are not fully established scientifically. In this context the proponent of an activity, rather than the public, bears the burden of proof.
>
> (Wingspread Statement 1998)

This version entails that action is based on hypothetical risks ("better to be safe than sorry") and it leads to bans or very restrictive regulations. The "weak" version, instead, states: "When human activities may lead to morally unacceptable harm that is scientifically plausible but uncertain, actions shall be taken to avoid or diminish that harm" (COMEST 2005). The Weak Precautionary Principle leads by default to public regulation that manages the potential risk.

In both cases, the Precautionary Principle shows: a) uncertain threat (knowledge condition): lack of scientific knowledge about the causal relationship between the action/inaction and the harm, that is, no conclusive evidence about the effects of the introduction of a new technology; b) possible harm as a consequence of a decision (damage condition); c) anticipatory action (action or remedy condition); d) burden of proof on those who introduce a new technology: technology should not be introduced until proven safe.

2.2 Cost-Benefit Analysis

The Cost-Benefit Analysis is based on the identification and calculation of benefits (positive impacts) and costs (negative impacts) of a technological innovation to determine if policy, program, or project will provide gains to society. Both benefits and costs are considered from a market, financial, and monetary viewpoint, and expressed in a common measure, typically dollars, including things not normally bought and sold on markets (for example, air is not marketed, but houses located in fresh air areas are; the risk of death is not marketed, but risky and non-risky jobs are) (Kelman 1981). Even social, ethical, and operational aspects are viewed from the market perspective.[1] Thus, when benefits overweight the costs, a technology can be introduced to the market. This principle allows the evaluation of actions in terms of maximization of benefits and minimization of harms, despite the action being safe, sustainable, ethically, or socially acceptable (Sagoff 1985).

2.3 Responsible Innovation

The Responsible Innovation consists of adopting a responsible view of innovation, where societal actors and innovators are mutually responsive to each other. They welcome innovation, try to anticipate potential risks of technologies, and work for social acceptance and sustainability of technologies in society (Valdivia and

Guston 2015; Koops et al. 2015). Innovation and technology should be "publicly acceptable, ethically unproblematic, and harmless" (Li et al. 2023): it should follow the principle of people-centered development and use to be compatible with respect to freedoms and human rights, including autonomy, dignity, privacy, and so on, ensuring they are ethically acceptable, then it should encourage a broad, democratic, and equitable implementation approach to avoid disadvantaging specific groups or individuals, and it must do no harm to human beings' physical and psychological health (Winfield and Jirotka 2018).

2.4 Proactionary Principle

The Proactionary Principle or Permissionless Innovation (first introduced by transhumanist philosopher Max More: see More 2004) is born in opposition to the Precautionary Principle, which is considered as stifling technological innovation, leading to bans or restrictive regulations and hindering effective competition of different players. This principle is in favor of the freedom to innovate without constraints. Innovation should be admitted by default, and only in case of serious harm (where there is a clear evidence of harm) it should be blocked as *extrema ratio*. Regulation should come in a responsive and flexible way and not as an anticipatory action. So, technologies should be considered safe, good, and healthy, unless and until it is shown to be otherwise. The assessment of risks should be based on science and not on popular opinion, emotions, etc. The burden of proof lies upon those who want to slow down innovation (Thierer 2014).

2.5 Innovation Principle

The Innovation Principle first made its appearance in a Staff Working Document of the European Commission's Directorate-General for Research and Innovation in 2016 (European Commission 2016), but it was launched in 2015 by an industry lobby group, namely the European Risk Forum (ERF), with support from Business Europe and the European Roundtable of Industrialists (ERT), whose members mainly come from the chemical, fossil fuel, and tobacco sectors. They insisted on the need for the EU to incorporate such principle "as an integral component of the policy-making process" (European Risk Forum 2015a). The idea is to complement the Precautionary Principle and to assess the impact of innovation, not irrespective of its effects on health and the environment, but "seek[ing] to support an evidence-based approach, based on the best available science, which is pragmatic, proportionate and renewable" (European Risk Forum 2015b). The purpose is to stimulate innovation by adopting a flexible regulation that takes into account the benefits and risks of alternative solutions too.

3 Methodology

Each of the aforementioned approaches has its strengths and weaknesses. In order to illustrate the strengths and weaknesses of the aforementioned approaches, a SWOT analysis approach (or SWOT matrix) was deployed. It is a project management technique, used by organizations to identify strengths, weaknesses, opportunities, and threats related to one project or business (Weihrich 1982). This methodology was selected because it provides a clear and structured framework to evaluate each approach based on its inherent characteristics and its suitability for addressing the complexities of AI governance. By applying the SWOT analysis, it was possible to assess the strengths of each approach, such as its ability to mitigate AI risks, promote innovation, or integrate ethical considerations, to examine the weaknesses, including any limitations or trade-offs associated with each approach. Additionally, the opportunities were explored, highlighting how each principle could contribute to advancing AI while minimizing harm, and the threats were identified. Ultimately, the SWOT analysis enabled a comparative analysis that is both systematic and comprehensive, allowing to highlight the key benefits and challenges of each approach in relation to AI governance. This comparison was crucial for determining which principles can provide the best balance between innovation and risk management in the evolving field of AI.

Table 4.1 summarizes the results of the SWOT analysis conducted on each of the five risk governance approaches:

From Table 4.1, it emerges that the "strong" version of the Precautionary Principle, the Cost-benefit Analysis *per se*, and the Proactionary or Permissionless Innovation Principle are not feasible and workable governance solutions for the following reasons:

- The Strong Precautionary Principle leads to blocking any technology and innovation because it states to avoid steps that will create a risk of harm ("better to be safe than sorry" principle): it aims for risk-zero (Sunstein 2002).
- The Cost-Benefit Analysis focuses too much on monetary values when some concepts are non-monetizable. It could become a technique which supports other principles, for example, supporting risk assessment and integrating other assessments where ethical, societal, and scientific elements are taken into account, but it cannot be a risk governance approach working alone and *per se*, because it is very partial and incomplete.
- The Proactionary or Permissionless Innovation principle leaves the technology uncontrolled and it may be too late to adopt measures for regulating a technology when harms to environment, health, etc. have already occurred.

Therefore, the Strong Precautionary Principle locks out AI innovation in health research and biobanking, and it should be discarded. Similarly, the Cost-benefit Analysis is focused on monetary values only, but many ethical, health, and social issues cannot be monetized, and thus this approach is incomplete for governing AI in health research and biobanking. Likewise, the Proactionary Principle would allow the AI to proceed uncontrolled and is thus not acceptable. For these reasons, these three approaches can be completely avoided for governing AI innovation.

The remaining approaches (i.e., the Weak Precautionary Principle, the Responsible Innovation, and the Innovation Principle) are more feasible as governance

Table 4.1 SWOT

Approach	*Strength*	*Weakness*	*Opportunity*	*Threat*
Strong Precautionary Principle	No strengths	Excessive focus on harms, neglecting benefits of a technology Focus on hypothetical risks	No opportunities	Blocking innovation, paralyzing principle, since it counsels to avoid steps that will create a risk of harm ("better to be safe than sorry" principle): it is not sustainable in real world
Weak Precautionary Principle	Science-based assessment of risks (i.e., scientific evidence of uncertainty is needed) Action should entail proportional measures ("acceptable risk") Motivating decision-makers in the public or private sector to explain and quantify their reasoning	It does not state what action to take: vague	Flexibility: actions based on the PP may be reviewed in light of new scientific data about risks Negotiable: it leads to discretionary actions for decision-makers Plural: potentially it might involve many stakeholders in the decision-making	Focus on worst-case scenarios may lead to bans or blocking regulation Excessive focus on quantitative risks, neglecting qualitative risks Regulation as a default position
Cost-Benefit Analysis	No strengths	It focuses on monetary/economic evaluation of costs and benefits of a technology, neglecting ethical, societal, etc. issues	No opportunities	Not everything can be calculated or evaluated in a monetary form
Responsible Innovation	Focus on responsibility, on public acceptance of a technology, on societal and ethical issues It considers innovation as a process and insists on the importance of involving many stakeholders and the public (participatory approach)	It risks being vague because regulation is not a default action Innovators have to prove the safety and security of technology but without constraints, so the burden of proof seems to be more on regulators	Flexibility of measures to be taken, which can be reviewed and adapted to circumstances	Involving too many stakeholders may lead to blocks to innovation because of incompatible views
Proactionary Principle	No strengths	It underestimates the unknown and hidden costs/risks of a technology	No opportunities	It leads to innovation and progress that proceeds uncontrolled: anarchy and "laissez-faire"
Innovation Principle	It aims to consider not only risks but also benefits of a technology: comprehensive assessment and science-based Proportional measures	Vagueness of types of actions to be taken	Flexible and adaptable Encouraging responsibility for innovation and transparency of decision-making processes	Risk of corporate interests prevailing over health, environment, safety, security, etc.

approaches for AI. However, each of them, if taken alone and on its own, is insufficient and inadequate.

Firstly, the Weak Precautionary Principle can be appreciated for the "scientific" approach for assessing risks, which shall be evaluated on the basis of scientific elements and not on the basis of mere hypothesis and irrational/emotional fears. Moreover, the fact of assigning the burden of proof to innovators appears reasonable, because it makes them accountable for the technologies they create, thus encouraging them to take into account not only the innovative elements but also safety, security, respect for environment, health, etc. However, the public regulation as a default position should be mitigated: since technology constantly evolves and changes, it is important that the regulation follows the same path, and sources of "hard law" and "top-down" complement "soft law" and "bottom-up" ones. Thus, regulatory measures should be subject to flexibility and review, in light of new scientific data and progress.

Secondly, the Responsible Innovation approach is helpful in stressing the importance of not considering scientific and quantitative aspects only, but taking into account societal, psychological, and ethical (i.e., nonquantitative and subjective) impacts of a new technology for a comprehensive risk assessment. Moreover, it insists on a participatory, multi-stakeholder approach, in order to better regulate a technology and reach a public social acceptance. So, the plurality element is a relevant one. However, putting the burden of proof on regulators would lead to decreasing responsibility of innovators, and this is risky.

Thirdly, the Innovation Principle is welcome with regard to the reference to proportionality measures, to flexibility and adaptability of measures according to the evolution of a technology: indeed, it calls for reviewing decisions if new knowledge or new uncertainties arise. It also focuses on a proper balance between progress and precaution. However, there is the risk of corporate interests prevailing over health, environment, safety, security, etc. and this principle is too vague in terms of actions to adopt.

4 Proposal for a New "Combined" Approach for Risk Governance of AI

Considering the features of the different approaches, it is argued here that the Strong Precautionary Principle, the Cost-Benefit Analysis, and the Proactionary Principle can be discarded for governing AI innovation. The other approaches (i.e., the Weak Precautionary Principle, the Responsible Innovation, and the Innovation Principle) are more acceptable, but they cannot be applied on their own, and thus a "combined" reading and approach is most suitable for risk management of AI applications. In a nutshell, such a "combined" approach consists of adopting some aspects of the Weak Precautionary Principle, the Responsible Innovation, and the Innovation Principle in conjunction as they complement each other for risk governance of AI.

More specifically, the Weak Precautionary Principle is helpful in order to: a) stress the need for a better scrutiny of the threats that AI may pose through a scientific-based risk assessment at early stage; b) pose the burden of proof and responsibility to the creators of algorithms and AI applications, since they have to explain algorithmic decision-making (especially in cases of "black box"); and c) guide the assessment of risks in cases of data processing in automated ways, that is, this principle is at the basis of the so-called Data Protection Impact Assessment (DPIA) provided by Article 35 of the General Data Protection Regulation (GDPR) for cases of data processing resulting in high risk to rights and freedoms of natural persons.

The Responsible Innovation approach may play a role in a) calling for a comprehensive risk assessment that takes into account environmental, societal, and ethical impacts of AI and b) calling for more transparent, participated, and more inclusive decisions on AI.

Thirdly, the Innovation Principle may encourage low-risk innovations and insists on flexible and constant evaluation – on a case-by-case basis – of the level of risk of AI applications. Indeed, the European Commission in its Communication on AI in 2018 (COM/2018/237 final) states,

> For any new regulatory proposals that shall be needed to address emerging issues resulting from AI and related technologies, the Commission applies the Innovation Principle, a set of tools and guidelines that was developed to ensure that all Commission initiatives are innovation friendly.

This principle underlines the importance of scrutinizing emerging technologies early in the legislative process.

Therefore, in our view, a proper risk governance of AI should be based on a combination of these three approaches that do not work as inflexible principles and static rules, but rather as a guide for decision-making. In this manner, the Precautionary Principle is not a stifle to innovation (Castro and McLaughlin 2019), but it can stimulate a sustainable innovation; the Responsible Innovation approach is not neglecting scientific and quantitative evaluations of risks, and the Innovation Principle is not a lobby instrument to "kill" any regulation, but a way for coupling innovation with environmental, safety, health, and rights protection.

So, our position is different from Gorgoni who suggests exclusively the Responsible Research and Innovation (RRI) model as a model for the responsible development of AI (Gorgoni 2020), or Hemphill who considers Responsible Innovation and Permissionless Innovation as providing the greatest potential for technological innovation to thrive (Hemphill 2020) and that better manage and regulate tangible risks of AI. Moreover, our perspective also goes beyond the one put forth by Rathenau Instituut (2020) about the combination of precaution and innovation in the event of uncertain risks. The attempt here is rather to combine three approaches, that is, the Precautionary Principle, the Responsible Innovation approach, and the Innovation Principle, for building a comprehensive and feasible model of risk governance of AI.

4.1 The EU AI Act and Its Governance Approach

The EU is the first entity in the world having set out a legal act on AI, which aims to settle harmonized rules for the development, placing on the market, and use of AI. The recently adopted Regulation is the world's first horizontal legislation on AI.[2] The text is the result of a long both political and legislative process comprising, among others, a White Paper on Artificial Intelligence[3] or public consultations.[4]

The rules establish obligations for AI, based on its potential risks and level of impact. The AI Act follows a risk-based approach, which "represents an attempt to face the new challenges of innovation through a rational and technocratic approach that fosters more efficient, objective, and fair governance, whilst fighting against 'over-regulation, legalistic and prescriptive rules, and the high costs of regulation'[(Macenaite 2017)]" (Dunn and De Gregorio 2022, 2). It aims to address the risks of specific uses of AI systems, categorizing them into four different degrees.

Some AI systems are outright prohibited because they pose an unacceptable risk to individuals or society. These include AI that manipulates human behavior, overrides free will, or exploits vulnerabilities due to age, disability, or economic status (European Commission 2023). Other AI systems fall into the high-risk category because they could significantly impact health, safety, fundamental rights, democracy, the rule of law, or the environment. Examples include AI used in medical devices or critical infrastructure like transportation and healthcare. These systems must meet strict standards for transparency, human oversight, accuracy, security, and risk management. They also require pre-market approval and registration in an EU database. AI systems with a limited risk face fewer restrictions but must still comply with transparency and traceability requirements, though they do not need formal approval. Finally, low-risk AI systems are not subject to specific legal requirements, but they are encouraged to follow best practices and ethical guidelines voluntarily.

Such a risk-based approach seems to be inspired by the Precautionary Principle. Through the prohibitions of unacceptable risk, "the AI Act adopts a precautionary approach, denying market access whenever the risks stemming from their use are deemed too high for risk-mitigating interventions" (Almada and Petit 2023, 8): for prohibited systems, the Strong Precautionary Principle has been followed.

When regulating high-risk AI systems – those that could significantly impact health, safety, and fundamental rights – the AI Act takes a Weak Precautionary Principle approach. This means that, rather than completely prohibiting such systems, the act imposes strict compliance obligations to minimize risks before and after deployment.

The AI Act requires providers and deployers of high-risk AI systems to take proactive measures. Before introducing these systems to the market, they have to identify and assess all known and foreseeable risks. Additionally, they are expected to continuously evaluate emerging risks based on real-world data gathered from the post-market monitoring system. Risk management is central to compliance. Organizations are required to adopt appropriate measures aligned with the latest technological standards, ensuring that any remaining risks are within acceptable limits. The overarching goal is to eliminate or reduce risks as much as possible through careful design and development. The Commission made reference to all the known and foreseeable risks and adopted a very strict standard ("risks be reduced or eliminated 'as far as possible' (AFAP) through design and development"): this has been criticized, since it is unclear what "as far as possible" means, which basically leads to indeterminacy of risk management (for a further analysis of this argument, see Fraser and Bello y Villarino 2021).

The Parliament's and Council's version instead requires a regular review and updating of the risk management process and requires to identify and estimate – more precisely – "the known and the reasonably foreseeable risks that the high-risk AI system can pose to the health or safety of natural persons, their fundamental rights including equal access and opportunities, democracy and rule of law or the environment" (Article 9(2)(a)). Plus, the risk management measures should be such that relevant residual risk is reasonably judged to be acceptable. Then, identified risks should be eliminated or reduced "as far as technically feasible" (Article 9(5)(a)) through adequate design and development of the high-risk AI system, involving when relevant, experts and external stakeholders. Therefore, the Parliament's and Council's version places more emphasis on reasonableness and proportionality in risk management. So, for high-risk AI systems, the Weak Precautionary Principle has been adopted.

Then, when the text makes reference to transparency obligations (Chapter IV of the AI Act) and recognizes the high level of protection to be ensured to fundamental rights (such as human dignity, respect for private life and protection of personal data, non-discrimination, and freedom of expression) through the provision of many proportionate obligations on all value chain participants (such as *ex ante* testing, risk management, and human oversight), the regulation looks to be in line with the Responsible Innovation approach (see also Articles 13 and 14 of the AI Act).

Finally, where the act encourages measures in support of innovation (Chapter VI of the AI Act), it seems to adopt the Innovation Principle: indeed, it gives national competent authorities the role

> to set up regulatory sandboxes and set a basic framework in terms of governance, supervision, and liability. AI regulatory sandboxes establish a controlled environment to test innovative technologies for a limited time on the basis of a testing plan agreed with the competent authorities.
>
> (European Commission 2021)

Technology companies can test and develop their innovations benefiting, for instance, from the exemption of the application of certain specific rules or legal regimes under a controlled environment.

In a nutshell, it seems that the EU has – maybe unconsciously – embraced a risk governance approach that somewhat combines the Weak Precautionary Principle, the Responsible Innovation, and the Innovation Principle. Yet, the EU AI

Act lacks some elements when it is read in the light of the proposed "combined" approach.

4.2 EU AI Act and Challenges of the "Combined" Approach

In the EU's AI Act, the prevalence of the architectural structure of the text seems to have been given to the Precautionary Principle (with reference to the burden of proof on innovators and to scientific-based risk assessments). However, some aspects should be reviewed in order to align with the approach suggested in this chapter: in particular, it would be better to clarify the criteria on the basis of which risk assessment is conducted. Indeed, in order to assess the types of risk that a certain AI application is causing, the AI Act should indicate the criteria, and not simply give a list of what categories of AI systems are deemed "unacceptable risk" and thus prohibited, otherwise it will be hard to argue which new systems would be added in future in the list without knowing the criteria for prohibition. Moreover, the burden of proof on innovators should be based on proportionality and reasonableness, and the idea of eliminating all the residual risks should be avoided, since it is based on a Strong Precautionary Principle which looks for an impossible "zero risk" situation. A proportional risk acceptability should instead be invoked.

When considering the Responsible Innovation approach, it appears that the AI Act does not sufficiently take into account environmental, societal, and ethical impacts of AI, when conducting the risk assessment of AI systems. Indeed, the text adopts a definition of "risk" (Article 3(2)) which means "the combination of the probability of an occurrence of harm and the severity of that harm". It "is characteristic of a product safety conception of risk, in which harm can be measured quantitatively" (Almada and Petit 2023, 19). It forgets that there are risks that cannot be measured, such as risks to fundamental rights caused by AI technologies. The AI Act should thus propose a more comprehensive view of risk assessment including evaluation of both quantitative and nonquantitative risks.

In addition, the AI Act forgets the role played by citizens who consume AI services, and thus the role of citizens should be boosted. Providers and deployers of AI systems must participate in assessing the risks, impacts, and potential harms of AI systems to both individuals and society, including impacts on fundamental/human rights. It means that those affected by the impacts of high-risk AI systems – both individuals and groups – must have input as stakeholders when those systems are certified as compliant (Edwards 2022). The act, instead, seems to give a role to developers of AI only.

It can also be observed that the EU AI Act follows a "top-down" approach, since it is the legislator that sets the threshold between acceptable and prohibited AI applications, and defines the different types of risks: indeed, "the list of 'unacceptable', and therefore prohibited, AI systems is directly set by the law and is independent of any a posteriori risk assessment by providers or users of those systems" (Dunn and De Gregorio 2022, 6). It would be better to modify such a perspective by more involving both the providers and deployers/users of AI with a "bottom-up" approach in combination with the "top-down" one. A proper "bottom-up" perspective can already be recognized in the role assigned to voluntary codes of conduct (industry-led ones) that can support the risk governance of AI; however, such codes are required only in cases of AI applications with minimal risks. It is recommendable to extend synergies with deployers and stakeholders.

4.3 Examples of Adoption of the "Combined" Approach in Health Research and Biobanking

After explaining the "combined" approach of risk governance for AI applications and the gaps in the AI Act in light of the "combined" approach, the further step consists of concretely applying such an approach to health research using AI (Paul et al. 2021). AI systems may accelerate and reduce the cost of discovering new drugs. Usually, a drug is based on synthesized molecules that aim to target the molecule involved in a disease. To find these molecules, research and tests are needed. AI can help in this regard screening libraries, analyzing, and identifying patterns in large datasets, predicting the behavior of a compound, or designing possible new drugs. However, uncertainties are related to the lack of *datasets having sufficient size and quality, the risk of black box medicine, etc.*

A combined approach offers a well-rounded strategy for governing AI-related risks by integrating different principles. The Weak Precautionary Principle ensures that AI research is grounded in solid scientific methods. When human rights and freedoms are at stake, this principle underpins the Data Protection Impact Assessment (DPIA), which must be conducted before processing data. The DPIA helps identify fundamental rights that could be affected and assesses potential risks during the development phase, allowing for the implementation of technical and organizational measures to mitigate those risks. The Responsible Innovation approach goes beyond technical assessments by incorporating societal and ethical considerations in AI research. It also emphasizes transparency and explainability as key tools for identifying and mitigating potential biases in AI systems. The Innovation Principle promotes proportionality, ensuring that research methods are appropriate for the intended goals. It also encourages an innovative application of the "privacy by design" approach, which requires researchers to prioritize data protection from the very beginning of their work, ensuring accountability throughout the research process.

Taking an example from biobanking, it is clear that AI "will have profound and positive impacts, with the overall outcome of increasing specimen/data utilization, greater repository sustainability, and increased speed of scientific discovery" (Grossman and Henderson 2023, 1). Let us consider a digital biobank for the development of clinical decision-making systems. The infrastructure collects, organizes, and uses clinical data through AI, machine, and deep learning techniques, so that massive amounts of structured and unstructured data are collected and processed quickly and simultaneously.[5] Beyond the "classical" risks highlighted in the literature relating to biobanks and referred both to the participant (or various stakeholders) and to the biobank itself (risks that might be physical, nonphysical, infrastructural, community, economic, societal, financial, etc.), the use of AI can lead to risks that, as observed by Akyüz et al. (2021, 20), "cannot only be thought of as material or physical, but may be embedded in values, practices, relationships, as well as broader phenomena, such as research integrity, representativity and societal justice".

So, when integrating AI into biobanking, a combined risk governance approach can help navigate the complexities of privacy, ethics, and innovation. The Precautionary Principle supports an adaptive risk assessment, as proposed by Akyüz et al. for biobanking in general – a concept that applies well to AI in biobanking. This principle acknowledges that risks are dynamic and context-dependent, evolving over time. By incorporating this flexible approach, AI governance in biobanking can continuously reassess risks as technology advances and new ethical concerns emerge. The Responsible Innovation approach encourages a holistic view of risk assessment, considering not just immediate technical risks but also broader societal and ethical implications. This means evaluating AI's potential long-term effects on patient rights, consent frameworks, and data integrity, ensuring that emerging risks are identified and addressed proactively. The Innovation Principle helps balance privacy risks with the societal benefits of scientific research and biobanking. AI has the potential to drive breakthroughs in medicine, but privacy concerns must be managed from the outset. By embedding privacy-by-design principles from the very beginning of AI adoption in biobanking, researchers can ensure that data protection is not an afterthought but an integral part of innovation.

5 Conclusion

AI has a lot of potential applications, and it can yield positive impacts for humanity; however, it can also give rise to unexpected and unintended consequences and determine new forms of risks that need to be effectively assessed. That is why the risk governance of AI, which is still neglected, constitutes an area of interest to which it is urgent to dedicate attention.

Ryan Budish (2021, 260) observes that, as of April 2020,

> there are, by one count, over 160 different sets of artificial intelligence (AI) governance principles, representing efforts from around the world, and from public and private organizations alike. What unites them is their attempt to grapple with the uncertain possibilities, both good and bad, latent within AI technologies.[6]

Finding an approach to govern risks that AI can generate is challenging.

Opting for quantitative models based on scientific assessment and calculation of risks only is not sufficient. On the other hand, qualitative models entirely based on societal, ethical, and political evaluation of risks excluding the scientific measurement of risks are not enough either. "Top-down" decisions that do not take into account and fail to involve different stakeholders from the "bottom-up" are also insufficient to address risk governance of AI.

AI in health research and biobanking presents unique challenges due to the sensitivity of health data, evolving risks, and the need to balance both innovation and ethical safeguards. This chapter suggests combining three approaches – the Precautionary Principle, Responsible Innovation, and the Innovation Principle – in order to adequately address these challenges. AI systems used in biobanking are not static – they evolve with new data, changing clinical insights, and advances in machine learning. The Precautionary Principle, particularly in its Weak version, ensures that governance remains dynamic and flexible. This means recognizing that risks are context-dependent and can shift over time due to new technologies or ethical concerns. It requires implementing continuous monitoring and reassessment to identify emerging risks before they become critical, allowing for regulatory flexibility so that safeguards evolve alongside AI advancements rather than becoming outdated.

AI in health research does not just present technical risks – it also raises societal, ethical, and legal concerns. The Responsible Innovation approach ensures, on the one hand, that risk assessment is holistic by extending risk evaluation beyond data accuracy and system performance to long-term effects on patient rights, consent frameworks, and equity in healthcare. On the other hand, it encourages transparency and explainability to mitigate biases, ensure fairness, and foster public trust while promoting stakeholder engagement by ensuring that researchers, policymakers, and the public play an active role in shaping AI governance (see also chapter "Artificial Intelligence and Communication With Research Participants in the Process of Biobanking Human Biological Material" by Jarosław Sak and Jakub Pawlikowski).

At the same time that safety and ethics are guaranteed, governance must preserve the potential of AI to drive medical breakthroughs (see also chapter "Powerful AI: Between Accountability and Potentiality" by Michaela Th. Mayrhofer). The Innovation Principle ensures that AI governance maintains proportionality, applying tailored safeguards that do not stifle beneficial research, and embeds privacy-by-design principles from the beginning. This establishes data protection as a core feature rather than an afterthought, while a structured but flexible regulatory environment is created that fosters both accountability and scientific progress.

Rather than treating risk governance as a rigid, one-size-fits-all model, this approach allows for nuanced, dynamic, and ethically sound oversight – guaranteeing that AI advances health research while protecting fundamental rights and public trust. The EU too seems to go toward this direction, having adopted a risk-based approach for its AI Act, which represents the first comprehensive legal instrument about AI established by a major legislator. This regulation constitutes a major step in the development of a regulatory framework for AI in Europe, providing a structured foundation for managing risks while enabling innovation. However, further refinements could enhance its effectiveness in health-related AI applications, by ensuring a more elaborated balance between trustworthy AI, risk mitigation, and the advancement of scientific progress in biobanking and medical research.

Acknowledgments

This chapter is the result of reflections and dialogue conducted within two European projects: Healthy Cloud (which has received funding from the European Union's Horizon 2020 research and innovation program under grant agreement No 965345), and the IMI project Bigpicture (grant agreement No 945358).

Notes

1 See also the chapter "Social Acceptance of Artificial Intelligence in Biobanking" by Elodie Caboux, Patrick Courtney, Io Hong Cheong, and Zisis Kozlakidis in this volume as regards financial aspects and market acceptance.
2 The AI Act was proposed by the European Commission on 21 April 2021 (see https://eur-lex.europa.eu/legal-content/EN/TXT/?uri=CELEX:52021PC0206). The Council of the European Union discussed the proposed act in October 2021. On 9 December 2023, negotiators in the Council and the European Parliament reached a provisional agreement on the AI Act. The final text passed the European Parliament in March 2024 and was unanimously approved by the Council on 21 May 2024, thus entering into force on 1 August 2024.
3 EU White Paper on AI: https://commission.europa.eu/publications/white-paper-artificial-intelligence-european-approach-excellence-and-trust_en (23.03.2025).
4 See public consultations on the EU's *Have Your Say* portal (e.g., European Commission 2020).
5 For example, see AIBiBANK project, Biobanking for Artificial Intelligence, www.aibibank.org/en (02.01.2024).
6 See also Algorithm Watch. "AI Ethics Guidelines Global Inventory". Last updated April 2020. https://inventory.algorithmwatch.org (02.01.2024).

Reference List

Akyüz, Kaya, Gauthier Chassang, Melanie Goisauf, Łukasz Kozera, Signe Mezinska, Olga Tzortzatou, and Michaela Th. Mayrhofer. 2021. "Biobanking and Risk Assessment: A Comprehensive Typology of Risks for an Adaptive Risk Governance." *Life Sciences, Society and Policy* 17 (1): 10. https://doi.org/10.1186/s40504-021-00117-7.

Almada, Marco, and Nicolas Petit. 2023. "The EU ACI Act: A Medley of Product Safety and Fundamental Rights?" *Working Paper RSC 2023/59*. Robert Schuman Centre for Advanced Studies.

Bajorek, Joan Palmiter. 2019. "Voice Recognition Still Has Significant Race and Gender Biases." *Harvard Business Review*, May 10. https://hbr.org/2019/05/voice-recognition-still-has-significant-race-and-gender-biases (10.02.2024).

Battineni, Gopi, Mohmmad Amran Hossain, Nalini Chintalapudi, and Francesco Amenta. 2022. "A Survey on the Role of Artificial Intelligence in Biobanking Studies: A Systematic Review." *Diagnostics* 12 (5): 1179. https://doi.org/10.3390/diagnostics12051179.

Budish, Ryan. 2021. "AI's Risky Business: Embracing Ambiguity in Managing the Risks of AI." *Journal of Business and Technology* 259.

Castro, Daniel, and Michael McLaughlin. 2019. "Ten Ways the Precautionary Principle Undermines Progress in Artificial Intelligence." *ITIF*, February 4. https://itif.org/publications/2019/02/04/ten-ways-precautionary-principle-undermines-progress-artificial-intelligence (10.03.2025).

Chukhnov, A., and Yuriy Ivanov. 2021. "Algorithms for Detecting and Preventing Attacks on Machine Learning Models in Cyber-Security Problems." *Journal of Physics: Conference Series* 2096: 012099. https://doi.org/10.1088/1742-6596/2096/1/012099.

Cihon, Peter, Matthijs M. Maas, and Luke Kemp. 2020. "Should Artificial Intelligence Governance Be Centralised? Design Lessons from History." *Proceedings of the AAAI/ACM Conference on AI, Ethics, and Society*, New York, NY.

CloudLIMS. 2022. "How Artificial Intelligence Is Transforming Clinical Research." June 30. https://cloudlims.com/blog/how-artificial-intelligence-is-transforming-biobanking.html (10.02.2024).

COMEST. 2005. *The Precautionary Principle*. New York: United Nations Educational, Scientific and Cultural Organization.

Craig, Lev, and Linda Tucci. *What Is Machine Learning? Guide, Definition and Examples*. Last modified August 2024. www.techtarget.com/searchenterpriseai/definition/machine-learning-ML (10.02.2025).

Crigger, Elliott, and Christopher Khoury. 2019. "Making Policy on Augmented Intelligence in Health Care." *AMA Journal of Ethics* 21 (2). https://doi.org/10.1001/amajethics.2019.188.

Dunn, Pietro, and Giovanni De Gregorio. 2022. "The Ambiguous Risk-Based Approach of the Artificial Intelligence Act: Links and Discrepancies with Other Union Strategies." *CEUR Workshop Proceedings*, 3221. https://ciencia.ucp.pt/en/publications/the-ambiguous-risk-based-approach-of-the-artificial-intelligence- (10.06.2025).

The Economic Times. 2023. "AI and Privacy: The Privacy Concerns Surrounding AI, Its Potential Impact on Personal Data." Last modified April 25. https://economictimes.indiatimes.com/news/how-to/ai-and-privacy-the-privacy-concerns-surrounding-ai-its-potential-impact-on-personal-data/articleshow/99738234.cms?utm_source=contentofinterest&utm_medium=text&utm_campaign=cppst (08.04.2025).

Edwards, Lilian. 2022. "Regulating AI in Europe: Four Problems and Four Solutions." Expert opinion, *Ada Lovelace Institute*. www.adalovelaceinstitute.org/wp-content/uploads/2022/03/Expert-opinion-Lilian-Edwards-Regulating-AI-in-Europe.pdf (02.01.2024).

European Commission. 2016. Better Regulations for Innovation-Driven Investment at EU Level: Commission Staff Working Document. Publications Office https://data.europa.eu/doi/10.2777/987880.

European Commission. 2020. Public Consultation on the AI White Paper, Final Report. https://ec.europa.eu/newsroom/dae/document.cfm?doc_id=68462 (08.04.2025).

European Commission. 2021. Staff Working Document, Impact Assessment report SWD (2021) 84, Accompanying the Proposal for a Regulation of the European Parliament and of the Council Laying Down Harmonised Rules on Artificial Intelligence (Artificial Intelligence Act).

European Commission. 2023. "Commission Welcomes Political Agreement on Artificial Intelligence Act." (Press Release) December 9. https://ec.europa.eu/commission/presscorner/detail/en/ip_23_6473 (02.01.2024).

European Commission. White Paper of the European Commission of 19 February 2020 on Artificial Intelligence – A European Approach to Excellence and Trust COM (2020) 65 final.

European Risk Forum. 2015a. *Better Framework for Innovation: Fuelling EU Policies with an Innovation Principle*. www.businesseurope.eu/sites/buseur/files/media/imported/2015-00536-E.pdf (02.01.2024).

European Risk Forum. 2015b. *Innovation Principle Q&A*. www.eriforum.eu/uploads/2/5/7/1/25710097/innovation_principle_q&a_5_march_2015.pdf (02.01.2024).

Ezziane, Zoheir. 2006. "Applications of Artificial Intelligence in Bioinformatics: A Review." *Expert Systems with Applications* 30 (1): 2–10. https://doi.org/10.1016/j.eswa.2005.09.042.

Frascarelli, Chiara, Giuseppina Bonizzi, Camilla Rosella Musico, et al. 2023. "Revolutionizing Cancer Research: The Impact of Artificial Intelligence in Digital Biobanking." *Journal of Personalized Medicine* 13 (9): 1390. https://doi.org/10.3390/jpm13091390.

Fraser, Henry L., and Jose-Miguel Bello y Villarino. 2021. "Where Residual Risks Reside: A Comparative Approach to Art 9(4) of the European Union's Proposed AI Regulation." *SSRN*. http://doi.org/10.2139/ssrn.3960461.

Gao, Xiaoquan, Sabriya Alam, Pengyi Shi, Franklin Dexter, and Nan Kong. 2023. "Interpretable Machine Learning Models for Hospital Readmission Prediction: A Two-Step Extracted Regression Tree Approach." *BMC Medical Informatics and Decision Making* 23 (1): 104. https://doi.org/10.1186/s12911-023-02193-5.

Gorgoni, Guido. 2020. "Stay Human. The Quest for Responsibility in the Algorithmic Society." *Journal of Ethics and Legal Technologies* 2 (1).

Grieger, Khara, Madison Horgan, and Christopher Cummings. 2023. "Let's Talk About Risk. A Guide to Identifying, Assessing, Managing, and Communicating Risk." *NC State Extension*, May 22. https://content.ces.ncsu.edu/show_ep3_pdf/1704468117/25387/ (08.04.2025).

Grossman, Gregory H., and Marianne K. Henderson. 2023. "Readiness for Artificial Intelligence in Biobanking." *Biopreservation and Biobanking* 21 (2): 119–20. https://doi.org/10.1089/bio.2023.29121.editorial.

Hemphill, Thomas A. 2020. "The Innovation Governance Dilemma: Alternatives to the Precautionary Principle." *Technology in Society* 63: 101381. https://doi.org/10.1016/j.techsoc.2020.101381.

Hossain, Elias, Rajib Rana, Niall Higgins, Jeffrey Soar, Prabal Datta Barua, Anthony R. Pisani, and Kathryn Turner. 2023. "Natural Language Processing in Electronic Health Records in Relation to Healthcare Decision-Making: A Systematic Review." *Computers in Biology and Medicine* 155: 106649. https://doi.org/10.1016/j.compbiomed.2023.106649.

Hussain, Wan, Wan Ishak and Siraj Fadzilah Siraj. 2002. "Artificial Intelligence in Medical Application: An Exploration." *Health Informatics Europe Journal* 16.

Kelman, Steven. 1981. "Cost-Benefit Analysis: An Ethical Critique." *AEI*, February 07. www.aei.org/articles/cost-benefit-analysis-an-ethical-critique/ (08.04.2025).

Kin, Yew Yong. "Five Main Challenges of Artificial Intelligent Technology." www.mdis.edu.sg/blog/five-main-challenges-of-artificial-intelligent-technology/ (02.01.2024).

Koops, Bert-Jaap, Ilse Oosterlaken, Henny Romijn, Tsjalling Swierstra, and Jeroen van den Hoven, eds. 2015. *Responsible Innovation 2: Concepts, Approaches, and Applications*. Cham: Springer.

Latonero, Mark. 2018. "Governing Artificial Intelligence: Upholding Human Rights & Dignity." https://datasociety.net/wp-content/uploads/2018/10/DataSociety_Governing_Artificial_Intelligence_Upholding_Human_Rights.pdf (02.01.2024).

Li, Wenda, Tan Yigitcanlar, Will Browne, and Alireza Nili. 2023. "The Making of Responsible Innovation and Technology: An Overview and Framework." *Smart Cities* 6 (4): 1996–2034.

Macenaite, Milda. 2017. "The 'Riskification' of European Data Protection Law through a Two-fold Shift." *European Journal of Risk Regulation* 8 (3): 506–40. https://doi.org/10.1017/err.2017.40.

Mesko, Bertalan. 2017. "The Role of Artificial Intelligence in Precision Medicine." *Expert Review of Precision Medicine and Drug Development* 2 (5): 239–41. https://doi.org/10.1080/23808993.2017.1380516.

Michailidis, Panagiotis, Athanasia Dimitriadou, Theophilos Papadimitriou, and Periklis Gogas. 2022. "Forecasting Hospital Readmissions with Machine Learning." *Healthcare (Basel)* 10 (6): 981. https://doi.org/10.3390/healthcare10060981.

More, Max. 2004. "The Proactionary Principle." *Extropy Institute*. www.extropy.org/proactionaryprinciple.htm (02.01.2024).

Müller, Vincent C. 2014. "Risks of General Artificial Intelligence." *Journal of Experimental & Theoretical Artificial Intelligence* 26 (3): 297–301. https://doi.org/10.1080/0952813X.2014.895110.

Paul, Debleena, Gaurav Sanap, Snehal Shenoy, Dnyaneshwar Kalyane, Kiran Kalia, and Rakesh K. Tekade. 2021. "Artificial Intelligence in Drug Discovery and Development." *Drug Discovery Today* 26 (1): 80–93. https://doi.org/10.1016/j.drudis.2020.10.010.

Price, W. Nicholson, and I. Glenn Cohen. 2019. "Privacy in the Age of Medical Big Data." *Nature Medicine* 25 (1): 37–43. https://doi.org/10.1038/s41591-018-0272-7.

Rathenau Instituut. 2020. "Recipes: A Study of Precaution and Innovation." May 19. www.rathenau.nl/en/knowledge-democracy/recipes-study-precaution-and-innovation (02.01.2024).

Rychnovská, Dagmar. 2021. "Anticipatory Governance in Biobanking: Security and Risk Management in Digital Health." *Science and Engineering Ethics* 27 (3): 30. https://doi.org/10.1007/s11948-021-00305-w.

Sagoff, Mark. 1985. "Risk-benefit Analysis in Decisions Concerning Public Safety and Health." *CSEP Module Series in Applied Ethics*. https://repository.iit.edu/islandora/object/islandora%3A10213 (08.04.2025).

Scherer, Matthew U. 2015. "Regulating Artificial Intelligence Systems: Risks, Challenges, Competencies and Strategies." *Harvard Journal of Law & Technology* 29 (2). https://jolt.law.harvard.edu/articles/pdf/v29/29HarvJLTech353.pdf (08.04.2025).

Sunstein, Cass R. 2002. "Beyond the Precautionary Principle." *University of Chicago Law School. John M. Olin Program in Law and Economics Working Paper*, 149. https://chicagounbound.uchicago.edu/law_and_economics/87/ (10.06.2025).

Tableau. 2024. "What Are the Risks of Artificial Intelligence (AI)?" www.tableau.com/data-insights/ai/risks (02.01.2024).

Taeihagh, Araz. 2021. "Governance of Artificial Intelligence." *Policy and Society* 40 (2). https://doi.org/10.1080/14494035.2021.1928377.

Thierer, Adam. 2014. *Permissionless Innovation: The Continuing Case for Comprehensive Technological Freedom*. Arlington, VA: Mercatus Center, George Mason University.

Valdivia, Walter D., and David H. Guston. 2015. "Responsible Innovation: A Primer for Policymakers." *Center for Technology Innovation*. www.brookings.edu/wp-content/uploads/2016/06/Valdivia-Guston_Responsible-Innovation_v9.pdf (08.04.2025).

Waring, Jonathan, Charlotta Lindvall, and Renato Umeton. 2020. "Automated Machine Learning: Review of the State-of-the-Art and Opportunities for Healthcare." *Artificial Intelligence in Medicine* 104: 101822. https://doi.org/10.1016/j.artmed.2020.101822.

Weihrich, Heinz. 1982. "The TOWS Matrix – A Tool for Situational Analysis." *Long Range Planning* 15 (2): 54–66. https://doi.org/10.1016/0024-6301(82)90120-0.

WHO. 2021. "WHO, Ethics and Governance of Artificial Intelligence for Health: WHO Guidance." https://iris.who.int/bitstream/handle/10665/341996/9789240029200-eng.pdf?sequence=1 (02.01.2024).

Winfield, Alan F. T., and Marina Jirotka. 2018. "Ethical Governance Is Essential to Building Trust in Robotics and Artificial Intelligence Systems." *Philosophical Transactions of the Royal Society A: Mathematical, Physical and Engineering Sciences* 376 (2133). https://doi.org/10.1098/rsta.2018.0085.

Wingspread Statement. 1998. "Science and Environmental Health Network, Wingspread Statement on the Precautionary Principle." Racine, WI. www.gdrc.org/u-gov/precaution-3.html (08.04.2025).

Yu, Ronald, and Gabriele Spina Alì. 2019. "What's Inside the Black Box? AI Challenges for Lawyers and Researchers." *Legal Information Management* 19 (1). https://doi.org/10.1017/S1472669619000021.

Zewe, Adam. 2021. "Toward a Smarter Electronic Health Record." *MIT News Office*, September 23. https://news.mit.edu/2021/medknowts-electronic-health-record-0923 (08.04.2025).

5 Artificial Intelligence, Sex, and Gender

Hypes, Hopes, and Potential Risks in Biobanking

Kaya Akyüz, Mónica Cano Abadía, and Melanie Goisauf

1 Introduction

> If somebody hides a thing behind a bush, seeks it again and finds it in the selfsame place, then there is not much to boast of, respecting this seeking and finding; thus, however, matters stand with the seeking and finding of 'truth' within the realm of reason. If I make the definition of the mammal and then declare after inspecting a camel, 'Behold a mammal,' then no doubt a truth is brought to light thereby, but it is of very limited value, I mean it is anthropomorphic through and through, and does not contain one single point which is 'true-in-itself,' real and universally valid, apart from man.
>
> (Nietzsche [1873] 1997, 93–4)

Nietzsche questions throughout his work well-established concepts in Western thought such as the Subject, the Self, Truth, Good, and Evil. These concepts, for him, are fictitious abstractions transformed into substances. One can only tell the truth about things in the specific sense and extent to which one can "find" something that one has previously constructed and hidden behind a bush. For Nietzsche, Truth has a discursive history that can be reconstructed by performing genealogies of knowledge, a method Foucault referred to as the archaeology of knowledge (Foucault 1977, 1972). Both authors seek to describe the conditions of the emergence of knowledge and discourses, highlighting the fact that there is no metaphysical substance that gives them the status of ahistorical, objective Truth. Neither knowing subjects nor the objects of knowledge are given independently of historically specific tactics and strategies that organize a broad spectrum of social practices. The production of "truth" is never entirely separable from technologies of power: "Truth is a thing of this world: it is produced only by virtue of multiple forms of constraint. And it induces regular effects of power" (Foucault 1980, 131). There is a discursive effort put into naturalizing certain ideas that are then presented as Truth: they are hidden behind a bush and then found again.

We chose to commence our critical examination of the convergence of research infrastructures, new data-driven technologies, and societal categorizations with this philosophical train of thought as it points to a fundamental characteristic of

DOI: 10.4324/9781032634487-8

the modes of scientific knowledge production: like Truth, "objective" knowledge is not simply there, it was constructed before (Knorr-Cetina 1981). In this chapter, we will show how the data-Truth gets "discovered" not only behind the bush but within the Big Data haystack. The continuous "rediscovery" of the "social" in the "scientific" has implications for doing science, especially biomedical research at a time when research has become so data-intensive and where biobanks as major research infrastructures are contributing to the knowledge production. Here, research is increasingly focusing on the status, definition, and understanding of data, due to their centrality for the knowledge produced. Scholars observe that raw data do not exist (Gitelman 2013) and data incorporate society's and researcher's assumptions while its production, appropriation, and use can be considered part of emerging social orders (Couldry and Mejias 2018). The status of data therefore needs to be regarded as situated, comprising inscribed assumptions about the social world and its organization, in the sense of Nietzsche's Truth that is not ahistorical or objective, but already constructed and rediscovered. Hence, data are situated, and social understandings of sex and gender are already part of theoretical assumptions and data practices, but those data practices and produced knowledge are equally important for understandings of sex and gender, in a co-productionist manner (Jasanoff 2004).

These considerations are currently of particular importance, as we see a hype around new data-driven technologies, hopes that they will improve healthcare, but also potential risks that they could perpetuate or even amplify injustices through biased datasets or data practices (see overview in Goisauf and Cano Abadía 2022) with lessons that can be learned from biobanking infrastructures regarding risks and opportunities (Akyüz et al. 2021, 2024). Big Data, machine learning (ML), and artificial intelligence (AI) are especially prominent in medicine and the life sciences as they are expected to contribute to diagnostics, prevention and therapeutics, efficiency, and accuracy with increasing amounts of data that would allow identification of minute details as patterns, as well as early or preemptive identification of disease conditions. These technologies encompass applications in screening and diagnosis, risk calculation and clinical decision-support, management planning, and precision medicine with the goal to improve patient outcomes, as well as decreasing costs, time, and labor. Such developments are driven by hopes that AI systems will reduce human bias and improve diagnostic performance, as, for example, in radiology, it has been shown that some could "outperform humans" (McKinney et al. 2020); however, the "outperformance" neither applies to the entirety of medicine nor is it realized in every AI application.

As with many new technologies, there is significant excitement and optimism surrounding big data, AI, and ML and their applications in biomedicine and biomedical knowledge production. While we recognize positive developments in terms of lowering sex bias, for example, in clinical trials, we highlight that the binary/gender bias is still prevalent even in calls against bias (Kim et al. 2010). Gender bias can lead to both false positives and false negatives, which can contribute to misdiagnoses, distress, and wrong and unnecessary treatments (UNESCO 2020, 7). AI systems have yet to be carefully examined so that adequate safeguards can be provided to protect research participants, data subjects, and users from harm. Therefore, we emphasize that the developments in AI and big data cannot be viewed as technical solutions for technical problems, but rather are enmeshed with choices. While we acknowledge that AI has great potential to benefit society, the risks during the data journeys (Leonelli 2020) in biomedical research and the data practices within the existing and emerging knowledge infrastructures, such as biobanks, need to be carefully considered (see also chapters "A Proposal for a Combined Risk Management Approach to AI Applications in Health Research and Biobanking" by Ilaria Colussi and "Liability for Damages, AI, and Machine Learning for Digital Pathology" by Dorota Krekora-Zając).

Building on existing literature on the matter and our own expertise in the field of biobanking, biomedical, as well as gender research (Akyüz et al. 2021; Goisauf et al. 2020; Goisauf and Cano Abadía 2022; Cano Abadía 2021; Delgado et al. 2024), we adopt a critical-interventionist approach to analyzing gender in the developments of Big Data-based research and AI applications in imaging and genomics. Our critical-theoretical discussion unfolds in the following manner: Firstly, we take a closer look at the increasing datafication in health research in connection with new technologies, in particular AI and ML; secondly, we examine the role of biobanks as major infrastructures in biomedical research; thirdly, we turn to sex and gender as key categories and potential sources of bias in situated, data- and technology-driven scientific practice, illustrating the claims with two cases; and finally, we discuss key takeaways from our critique and conclude with further directions for intervention.

2 Data, Health, and Algorithms: Hopes and Hypes of Precision

Scholars observe that there is an increased "technoscientization of biomedicine" (Clarke et al. 2003). One salient aspect of this process is the increasing scope and quantity of data in knowledge production, from genome-wide association studies of millions to large-scale analysis of electronic health records. However, knowledge production is also tied to technologies of automation, from risk analysis to quality metrics in the healthcare setting, as well as expansion of the clinic into the individuals' homes and daily lives, from tracking tools to telemedicine (Clarke et al. 2021, 132–4). These developments bring with them the expectation that more data will result in better understanding of differences in health conditions and their outcomes.

Biomedical developments from xenotransplantation to pharmacogenomics have been part of generations of expectations and hope that such innovations will "substitute or replace existing ways of doing things" (Brown 2003, 11; Geels and Smit 2000) and are "productive" (Adams et al. 2009, 257), as they contribute to new ways of generating (bio)value out of life (Rose 2007, 6–7). For instance, increasing data and developments in genome-wide association studies are expected to allow identification of new genetic variants, disease genes, rare variants, copy number variations, better understanding of biological pathways, development of novel clinical applications, and explaining differences between populations among others (Tam et al. 2019). The increase in data is at the same time tied to the use and development of technologies from preimplantation genetics to polygenic risk scores, which may impact different groups of individuals differently.

Alongside or as part of these developments, there are large-scale projects and infrastructuring efforts, identified as "the ongoing and continual processes of creating and enacting information infrastructures" (Karasti and Blomberg 2018, 2). An example of this could be the *All of Us Research Program*[1] in the United States, which was previously called *Precision Medicine Initiative*. At the time of its launch, some of the expected "near-term successes" were described as follows: "insights into pharmacogenomics – enabling the provision of the right drug at the right dose to the right patient", identification of "attractive drug targets for broad patient populations", as well as improvement of "strategies for preventing and managing chronic diseases" (Collins and Varmus 2015, 795). During the seven years that have gone by since the launch, the initiative has been transformed in multiple ways with changing expectations of what it can accomplish and the major name change to "All of Us" is a signal of how the project aims to attract people from diverse backgrounds, but also that participants' "[d]ata will be accessed broadly for research purposes".[2] In this regard, increased data both in scope and amount as well as the maximization of its use in research are tied to the expectation that intense datafication will lead to precision and thus contribute to improved health, supported by systems that are amenable to automation, increased efficiency, and productivity, generating a promissory discourse (Hoeyer 2019). Scholars note the need for a realistic assessment and acknowledgment of structural discrimination as the first step of "leaving no one behind" in the future health and precision medicine initiatives (Geneviève et al. 2020). Similar projects also exist at the European level, such as ICPerMed[3] and 1+ Million Genome Initiative,[4] as well as at the national level, for example, NHS Genomic Medicine[5] in the UK.

As part of both knowledge production in biomedicine and health services, there is much hope and hype about the potential of AI (Fröhlich et al. 2018; Davenport and Kalakota 2019). The hype and hopes about (future) potentials of AI systems must be put in relation to current challenges, particularly in design and application, with potential risks and negative societal impacts on vulnerable groups, but also in view of the vulnerabilities they may create. Data used to train AI and to build algorithms carry more than medical information; they may also reflect inequalities in societies that become "inscribed" (Akrich 1992) in algorithms and, thus, embedded in AI systems. Scholars have pointed out that unequal access to medical services may result in bias in the AI algorithms, where belonging to certain social categories might influence the diagnosis despite no direct link to the disease (Obermeyer et al. 2019; Ghassemi and Nsoesie 2022). In this regard, existing inequalities may be exacerbated through technological systems that may even be more difficult to correct than the current ones.

Algorithms and technologies based on AI and ML have the potential of spreading and reinforcing harmful gender, racial, and other stereotypes since the imaging, pathological, genomic, and clinical data that are feeding these technologies are linked to information that is reflective of societal dynamics. In this regard, bias can be introduced at many different stages of the development of AI techniques (e.g., generation, collection, and labelling of the data that go into the datasets; and determination of which datasets, variables, and rules the algorithms learn from) and become embedded in AI systems. While bias and discrimination have always existed, the issue nowadays is that AI and ML could both amplify and perpetuate their impact (Coeckelbergh 2020). The current scientific landscape turns the issue of bias into both an ethical and societal issue that is specific to data science-based AI, which can benefit from analyses of discrimination that have been and are being done in other disciplines. Different projects propose attending to bias from the beginning and having an ethics-by-design approach,[6] while the flourishing infrastructures, like biobanks, allow rethinking of data practices, from inclusion of categories to development of quality markers.

3 Infrastructures: Biobanks and Data

As discussed before, medical research and healthcare have become data-intensive and large amounts of health data are increasingly produced, (re)used, and shared, where large-scale infrastructures play a central role. The diversity of such infrastructures is immense, from genome sequences of hundreds of organisms[7] to databases of clinically relevant human genetic variations.[8] Along with these tool-like infrastructures, biobanks are becoming central to the knowledge production practices as they allow access to biosamples and data.

Considering that AI and ML technologies mainly rely on big data, it becomes important to understand where the data are coming from. Often, such big data are not produced for the purposes of a specific research project but rather for recreational purposes (e.g., genetic ancestry databases of private direct-to-consumer companies), may be pooled from different sources (e.g., platforms collecting images from the web), may originate as part of medical services (e.g., electronic health records), or may be collected by biobanks among others. In this chapter, our interest lies in biobanks as one of the infrastructures that provide the data that AI/ML is utilizing.

Kaye (2011) noted more than a decade ago that there is a need for governance that can manage the data flows of a globalized and interconnected system of biobanks, in comparison to the previous model of knowledge production where localized, small-scale, and project-based data production was at the core. As we strive toward open access, the valuation of data is also tied to its findability, accessibility, interoperability, and reusability, known as FAIR principles (Wilkinson et al. 2016). Pursuit for all encompassing data, however, does not mean that certain data are not excluded at different stages, from generation of data to analysis, where data are tailored according to the needs of the researcher. In this regard, FAIR does not always mean fair to individuals, groups, and their health, about which knowledge is produced using the data (Holub et al. 2018; Murtagh et al. 2021) (see chapter "Fairness of an AI System in the Case of a Biobank of Images and Imaging Biomarkers" by Valentina Colcelli).

Biobanks are one of the major infrastructures for biomedical research and a site for social scientists and philosophers in analyzing how and which epistemic values guide "biobanking science" (Demir and Murtagh 2013) and the data practices involved. The term biobank may be new; however, the practice of the storage of samples and data in biomedical research is quite old and the recent transformation is more in the openness of the infrastructure toward experimentation while the accompanying industrialization of the processes are bringing a new significance and contingent meaning to materials and data (Mayrhofer 2013). Despite the expectations and hopes of objectivity and completeness, data, regardless of their size and scope, are "inherently partial, selective and representative" (Kitchin 2014, 3). In this sense, as Bowker and Star (2000) note, neither data nor data infrastructures are neutral (p. 24) and choices are made in choosing data and building such infrastructures (p. 5). While noting the contingency and context-dependence (Mayrhofer 2013), these choices may on the one hand relate to the epistemic purpose of the biobank; for instance, as Demir and Murtagh (2013) note "reusability" and "re-purposability" are central to the epistemic choices made in infrastructuring biobanks. However, these choices extend to the level of data itself. In this regard, we concur with scholars who argue that "raw data is an oxymoron" (Bowker 2005, 184; Gitelman 2013) and that "databases and data infrastructures are not simply neutral" (Kitchin 2014, 21). We will consider these issues in the next section through a specific topic, sex/gender, and explain why it is important to think about big data and AI together.

4 Sex and Gender, Big Data, and AI: From Precision to Exclusion

Ludwik Fleck notes, "Whatever is known has always seemed systematic, proven, applicable, and evident to the knower" (1981 [1935], 22). In this regard, like other scholars interested in the entanglements of science, technology, and society, Fleck highlights the importance of critically approaching produced knowledge. Taken together, these theorists point out that we cannot assume truth as simply given – or in that sense objectivity in scientific knowledge – but that knowledge must be understood as constructed and embedded in social and power-pervaded structures, that is, that knowledge is situated (Haraway 1988). Looking at sex and gender, the history of medicine has shown that the understanding of illness and the body was – and is – sexed and gendered (Laqueur 1992). For example, there is a history of inequality in medicine, as male bodies were long seen as the standard in medical trials, whereas female and gender-nonconforming bodies were missing (Kacel et al. 2022). In the last few decades, there have been efforts of "niche standardization" to transform categories of humans into distinct populations, for which a standard body is then imagined (Epstein 2007, 135). Epstein notes that this approach emerged as "an alternative to either complete generality (a one-size-fits-all embrace of the 'standard human') or full particularity (a focus on the individual)" (p. 278). With the emergence of big datasets, algorithms, and medical AI, the imaginaries of "standards" and "standardized bodies", and the categories that are used and ignored become ever more important for the analysis of sex and gender bias.

Humans are involved in many stages of data practices, and they carry their human knowledge and their conscious and unconscious biases with them. In this sense, and paraphrasing Haraway (1988), it is important to talk about situated data (Rettberg 2020): data that are embedded and intertwined with particular historical, socioeconomic conditions, and technologies of power. In the words of one of the pioneering scholars in feminist AI: "It does become important to ask how AI systems are used to represent knowledge, what kind of knowledge and whose knowledges they contain" (Adam 1995, 356).

In this sense and adhering to the notion that there is no raw and neutral data, we argue that there are theories of gender that have an impact on which data are collected and how. Our current sex/gender system (Rubin 1975), which is binary and hierarchical, gets embedded in the data that are collected. Integrating a perspective that interrogates the kind of knowledge that AI systems represent requires us to rethink the binary system of gender and sex, along with other systems of classification that perpetuate oppression (D'Ignazio and Klein 2020). In this regard, as if echoing Nietzsche's metaphor of the bush, the classification systems, the gender and sex binaries being two of them, are constructed in a particular context and then they are used to sustain their relevance.

The usual categories of gender that are taken into account are women and men, but there are other gender categories such as agender, non-binary, or gender fluid who are dropped from studies because their sex and gender do not match the conventional understanding of the sex/gender system (Ganna et al. 2019). The categories that are not counted (like trans, non-binary, or agender people) become invisible along with individuals identifying as such. It is important to collect data that reflect the population that it purports to represent (D'Ignazio and Klein 2020, 98) and there are studies that highlight the importance of integrating transgender, non-binary, and gender nonconforming people in life sciences research (DuBois and Shattuck-Heidorn 2021; Ritz and Greaves 2022; Maas et al. 2024) and biobanking (Jones, Reyes, et al. 2020; Jones, Otto, et al. 2020; Ackley et al. 2023).

Both gender and sex are social constructs. Regarding gender, one of the most important contemporary theories explaining its mechanisms of formation and reproduction is Judith Butler's theory of gender performativity. For Butler, gender is an iterative process, a repeated performance, and a collection of performative acts sustained over time (Butler 1990). We all constantly repeat and perform gender norms to conform to what we are taught is "normal" behavior. Through repetition and naturalization, gender categories become reinforced: they are presented to us as the natural way to be. Gender norms establish what counts as "normal" gendered behavior (Butler 1996, 3).

Sex is also a social construct, according to Butler (1996, 2–15) and other feminist scholars (Fausto-Sterling 2012; Martin 2001). The origin of the current binary sex system can be traced back to the European Enlightenment (Laqueur 1992) and there have been different classifications throughout history, such as Linnaeus' still-in-use binomial classification of sex as well as his now-abandoned classification of humans into five "racial" subcategories. Feminist biologist Anne Fausto-Sterling, whose work has been dedicated in the last decades to the deconstruction of the sex binary (Blackless et al. 2000; Fausto-Sterling 2000), claims, "There is no single biological measure that unassailably places each and every human into one of two categories – male or female" (Fausto-Sterling 2018). Recently, to show the staggering complexity of the human sex spectrum, Amanda Montañez (2017) has created a complex diagram to visualize it beyond the binary.

Sexuality, or more concretely, heteronormativity (Warner 1991) is also part of this norm. We are taught what normal sexual-affective relationships are through what Butler calls the heterosexual matrix: a "grid of cultural intelligibility through which bodies, genders, and desires are naturalized" (Butler 1990, 194), a matrix of intelligibility that only presents us with one "natural" way of being and feeling. Heteronormativity assumes the sex-gender binary, considers that the only possible affective and sexual relationships are between people of the "opposite" gender, and shapes our understanding of social relations (Ahmed 2020; Sedgwick 2008). Gender performativity contests the essentialist accounts on sex, gender, and sexuality that consider that there are innate, natural, and essential criteria that make one heterosexual or gay, a woman or a man, and invites us to pay attention to the social mechanisms that produce and sustain the current sex/gender system. People who do not perform well within the norms of the sex/gender system (i.e., people whose gender and sex "do not match" with each other or people with genders beyond the binary) are beyond the limits of normativity and intelligibility (Butler 1996, 3). In this sense, Butler asks: What bodies matter? Whose lives count as important?

In relation to the performative perspective, AI depends on normativity and it has effects on the materialization and reification of identities (Keyes et al. 2021). In the case of sex and gender, this may lead the AI systems to fail to recognize identities that are trans, intersex, or non-binary since the people not belonging to the established normative categories remain outside of the frames of understanding of AI. In this regard, individuals exist as bodies and in the medical system become patients, get diagnosed, and are treated for a variety of conditions, but do not exist in data or are sometimes erased for various reasons as will be discussed shortly. This amounts to what Butler calls "performative contradiction" (Meijer and Prins 1998, 280) in that the existing bodies become excluded individuals not only from data but the discourse, leading to a participation problem. In health, this may mean, just as the past practices of excluding women from clinical research had severe consequences, the lack of data and knowledge acts against the individuals, who do not belong to the normative categories. In line with a more inclusive, sex- and gender-sensitive perspective, The American Medical Association updated its policies in 2018 to affirm the medical spectrum of sex and gender, by stating,

> Sex and gender are more complex than previously assumed. . . . It is essential to acknowledge that an individual's gender identity may not align with the sex assigned to them at birth. A narrow limit on the definition of sex would have public health consequences for the transgender population and individuals born with differences in sexual differentiation, also known as intersex traits.
>
> (American Medical Association 2018)

Sex, gender, and sexuality, as social categories of the sex/gender system, aim at classifying people within mainly binary structures that naturalize certain behaviors, desires, attitudes, and ways of being. Aiming at producing knowledge that is inclusive and that reflects the world we live in, it is crucial to pay attention to the mechanisms of formation of the sex/gender system and how science and scientific knowledge are imbued with these structures. It is important to analyze databases and data collection practices as situated in contexts that are not neutral and universal. Classification systems (sex, gender, and sexuality being some of the categories used to fix our identities) have a tendency of reproducing power imbalances of those who contribute to sustaining the stability of these systems (Noble 2018, 137; McQuillan 2022).

To illustrate these considerations and to show how social categories such as sex/gender are (re)constructed in and through datasets, get mobilized within (data) infrastructures, and are utilized for the development of new data-driven technologies, we will briefly introduce two cases: first on genetics of same-sex sexual behavior, and second, on imaging.

4.1 Case 1: The (Non-) Simplification of Same-Sex Sexual Behavior

The first case is a recent large-scale genome-wide association study (GWAS) that aimed to identify the genetic basis of same-sex sexual behavior (Ganna et al. 2019), and a critical discussion of this research by Goisauf et al. (2020). The sample compiled for the study consists of data from biobanks, which were collected with different methods.

The study exemplifies how social categories that are largely shaped by societal understandings of difference and order get reproduced, how the complexity of the social is reduced through specific data practices, and how social categories and knowledge production converge in data-intensive research. This can be traced in some simplifying methodological limitations, for instance, by building on heteronormativity and gender binary, manifested in the exclusion of noncomplying individuals from the study. Thus, even in the largest genome-wide association study on sexual orientation, certain categories that are relevant are excluded, while others are flattened.

The case reflects understandings of (genomic) data and (big) data practices. Categories used to study genetic variation and to advance personalized medicine rely on traditional epidemiological categories and individual characteristics, such as differences in gender, ethnicity, educational and socioeconomic status that are reinscribed in DNA (Prainsack 2015). The use of these categories often neglects that they have a political and cultural meaning within the social world that is structured by power relations. From this perspective, "raw" data do not exist but rather data are situated within specific contexts and produced within theoretical framings and methodological decisions (Kitchin 2014).

With this case, we want to highlight the potential consequences of such a simplification of complex social categories, relations, and phenomena while processing big datasets. As Coeckelbergh (2020, 91) states, "abstraction from reality is never neutral". Thus, the results can only partially depict the complexity of social reality, but can nevertheless be incorporated into the basic assumptions of further data-based research and development.

4.2 Case 2: From More Data to Excluding Data Practices

The previous case resonates with the notion that better, that is, bigger, data would yield better outcomes. However, it also highlights that data are co-produced with social and cultural assumptions, which means that no data can be viewed as "objective" or a source of "the truth" from this perspective. Such datasets are used to develop new data-based technologies, such as medical AI. While large amounts of data are needed to train algorithms, in this case images for radiological AI, the data that are used to train AI represent a key challenge. Data can be biased by comorbidities, by the labels given to the data which can contain human values and beliefs, or by differences in data caused by using different scanning techniques or machines (Geis et al. 2019), which calls into question the "objectivity" of the outcomes. Quinn et al. (2022, 4) point out that

> most training data are imperfect because learning is done with the data one has, not the sufficiently representative, rich, and accurately labeled data one wants. . . . even a theoretically fair model can be biased in practice due to how it interacts with the larger healthcare system.

Scholars are noting the increasing importance of synthetic data in AI (Floridi 2020) and there is debate about the potential of synthetic data to benefit the underrepresented groups in AI, but such expectations are also accompanied with risks and technical difficulties for the time being (Chen et al. 2021). Finally, it is difficult to neatly separate research and applications of AI in biomedicine and healthcare settings and thus the implications are best thought of as co-constitutive.

Datasets pursuing normative understandings of sex and gender or normalizing classifications to reduce noise in AI may risk reproducing or even amplifying existing gender inequalities. Ignoring the sex and gender dimensions and their contribution to health and disease may lead to discrimination (Cirillo et al. 2020). The use of gender-imbalanced datasets (Larrazabal et al. 2020) may have discriminatory effects and may have affected the performance in pathology classifications with chest X-rays of minority groups. Using more gender-balanced data, AI-driven systems were able to detect diseases more accurately.

In the discourse on ethics of medical AI, several societal implications are rather underexplored, requiring more attention on potential discriminatory effects and a stronger social science perspective in AI developments (Goisauf and Cano Abadía 2022). There is a risk that AI systems might have sex/gender bias as well as other biases, such as racial, which are introduced through the data used to train algorithms. As in other areas in society, sex and gender bias in AI technologies could have a negative impact on vulnerable groups and could result in discrimination in the healthcare system and beyond.

The assumption of "objective" outcomes must be rejected due to datasets that may carry human values and social structures. As we have seen in the previous case, narrow, that is, binary, assumptions used in data-intensive research and development regarding sex and gender, could lead to an underrepresentation of intersex, transgender, non-binary, or gender nonconforming individuals and, thus, may result in biased AI applications that are reproducing these assumptions in the long run.

5 Discussion

In this chapter, we have shown how knowledge is situated and how the mode of production of contemporary knowledge is shifting to a datafied terrain. In this terrain, biobanks are central, first, as entities that ease the sample and data circulation. However, at the same time, they serve a critical role in controlling, harmonizing, expanding, and limiting which kinds of data/samples are to be collected and which categories are going to be built into the infrastructures. Considering the imaginaries of precision and efficiency that AI brings into biomedical research, we highlighted what we can learn from social sciences, such as performativity, to expand the narrow understandings of sex and gender that are part and parcel of the current knowledge production practices, and potential future uses of AI.

As this chapter is an intervention, we wanted to collect risks, but also identify opportunities from our discussion of sex/gender, AI, and biobanking. In this regard, we summarize our key takeaways as follows, as a yet incomplete list of proposals.

5.1 Exclusion Has Consequences

It may be argued that individuals are different in so many ways that it is not possible to produce biomedical knowledge without categorically analyzing them and thus certain categories will be more prominent while others will not receive the same attention, often due to size. Following the devastating consequences on the fetus of drugs like diethylstilbestrol and thalidomide used for morning sickness without prior testing on pregnant women, in countries like the United States, women with childbearing potential were excluded from clinical trials for decades in a paternalistic approach (Macklin 2010; Corrigan 2002). Taking history as a yardstick, it is clear that exercises of exclusion had serious consequences, from individual health to social justice. However, such exclusionary practices also have consequences in science and inclusion is thus not merely a social justice issue. Efforts in genetics toward understanding the etiology of rare diseases are an example of how despite the small size of those who are affected, the findings from such research can benefit the global population because knowledge generated allows new insights into biological mechanisms (Posey et al. 2019).

5.2 We Need Representation, Not Flattening of Individuals Into Standardized Humans

Issues of representation are often considered through the critique of European/ Western-focused knowledge production practices, where certain populations are underrepresented in the knowledge production process due to a variety of reasons, often historical inequalities as well as mistrust. While there is a need to remedy the representational bias by including the entire diversity of the human population from geography to gender, we caution against the potential risk of creating new "standard humans" for each category, in ways that individuals belonging to underrepresented groups are then flattened into new, but multiple standard humans. This practice, a kind of "niche standardization" as Epstein (2007) puts it, should not be the solution between an old imaginary of a standard human (often in the form of a white, European, heterosexual man) and the imaginary that each individual is different. Instead, lowering representational bias should be a prerequisite for more inclusion.

5.3 Representation Is Not Enough: Inclusion Has to Be Democratic

As the previous point on representation highlights, the samples and data collected may be as representative of the entire global population from gender to geography as possible. However, as we have critically approached truth and objectivity along the lines of scholars before us, we highlight that knowledge is not produced in a vacuum, but is situated in practices of scientists, from deciding which data are worth collecting to coding data with identifiers. As has been argued by scholars on big data (Zook et al. 2017) and anti-racism (Chaudhary and Berhe 2020), there is a need for societal debate and the inclusion of actors from underrepresented communities in the research process, from recruiting researchers belonging to underrepresented groups to open discussions. Nonetheless, the inclusion of diverse populations needs to be meaningful and focuses on creating environments where diverse voices are not only present but also valued and empowered; this would allow us to escape "the paradox of exposure: the double bind that places those who stand to significantly gain from being counted in the most danger from that same counting (or classifying) act" (D'Ignazio and Klein 2020, 105). It is important to establish methods to avoid marginalized people feeling tokenized, which often happens when being included without addressing underlying and systemic issues of discrimination. With AI, it becomes evident that producing algorithms that are explainable (Rasheed et al. 2022; Brady and Neri 2020) is an important area of research; however, considering the potential impacts of these technologies, rather than communication about the final products, engagements during the process should be the goal to make inclusion democratic.

5.4 Potential for More, Better Knowledge: More Data Is Not Better Data

Some authors highlight the importance of introducing desirable bias to counter undesirable bias: "desirable bias implies taking into account sex and gender differences to make a precise diagnosis and recommend a tailored and more effective treatment for each individual" (Cirillo et al. 2020, 2). It is crucial to increase awareness of unintended biases among the scientific community and involve other stakeholders to understand where bias that may potentially lead to discrimination might be introduced. In this regard, while we acknowledge data used for an AI algorithm may reproduce the bias in the original sample, we agree with Epstein, that "biomedicine is a domain in which conceptions of identity and difference actively get worked out in practice" (Epstein 2007, 16) and thus we cannot merely rely on more data being the solution.

A series of questions can be asked to analyze the social implications of AI systems (Schiebinger and Klinge 2020): Who will benefit? Who will be left out and/or put at risk? Are background assumptions about sex and gender shaping or embedded in the concepts and theories of the field? Do researchers assume that sex and gender are fixed and binary? Considering these types of questions during all stages of research, creation, and implementation of AI systems may raise awareness on the social implications of these systems and help correct some of the potential discriminatory results.

5.5 New Roles for Biobanks: Capacity for Making a Difference

While biobanks serve as "custodians" of samples and data and thus seem to be serving a passive, but supportive role in knowledge production, as shown before, they are infrastructures in which certain choices and values are inscribed. This gives biobanks as infrastructures the potential to make a change. International quality standards (ISO 2018, 2012; ISO and IEC 2013) through which institutions may get certified or technical standards such as "Minimum Information About BIobank Data Sharing" (Merino-Martinez et al. 2016) that are created for compatibility are examples of what biobanks often consider from the harmonization perspective. With debates about sex/gender, we highlight that just as quality markers can be given for samples, biobanks and consortia of such infrastructures may push efforts to produce quality markers for sex/gender for samples and data.

6 Conclusion

Based on the literature on sex/gender, big data, and AI, the two cases that we have discussed led us to these five propositions that can be explored further, just as other ones can be contributed by other scholars. In a field that is experiencing both rapid growth and transformation, we highlight the necessity to think outside the box, critically view what is taken for granted, and engage in continuous debate not only within the research community or academic circles but also with the society and foremost those who are and will be impacted by the knowledge and technologies produced.

Disclaimer

Where authors are identified as personnel of the Biobanking and BioMolecular resources Research Infrastructure (BBMRI-ERIC), the authors alone are responsible for the views expressed in this chapter and they do not necessarily represent the decisions, policies, or views of BBMRI-ERIC.

Acknowledgments

Activities conducted under EOSC-Life, a project that has received funding from European Union's H2020 program (Grant agreement ID: 824087), contributed to the writing of this chapter.

Notes

1 *All of Us* Research Program: https://allofus.nih.gov/ (24.06.2022).
2 See the announcement on name change: www.nih.gov/allofus-research-program/pmi-cohort-program-announces-new-name-all-us-research-program (24.06.2022).
3 ICPerMed International Consortium: www.icpermed.eu/ (24.06.2022).
4 European "1+ Million Genome" Initiative: https://digital-strategy.ec.europa.eu/en/policies/1-million-genomes (24.06.2022).
5 NHS Genomic Medicine Service: www.england.nhs.uk/genomics/nhs-genomic-med-service/ (24.06.2022).
6 See, for instance, the European Commission report *Ethics By Design and Ethics of Use Approaches for Artificial Intelligence* (2021): https://ec.europa.eu/info/funding-tenders/opportunities/docs/2021-2027/horizon/guidance/ethics-by-design-and-ethics-of-use-approaches-for-artificial-intelligence_he_en.pdf (24.07.2023).
7 Ensembl Species List: www.ensembl.org/info/about/species.html (02.02.2022).
8 ClinVar: www.ncbi.nlm.nih.gov/clinvar/ (02.02.2022).

Reference List

Ackley, Sarah F., Scott C. Zimmerman, Jason D. Flatt, Alicia R. Riley, Jae Sevelius, and Kate A. Duchowny. 2023. "Discordance in Chromosomal and Self-Reported Sex in the UK Biobank: Implications for Transgender- and Intersex-Inclusive Data Collection." *Proceedings of the National Academy of Sciences* 120 (18): e2218700120. https://doi.org/10.1073/pnas.2218700120.

Adam, Alison. 1995. "A Feminist Critique of Artificial Intelligence." *European Journal of Women's Studies* 2 (3): 355–77. https://doi.org/10.1177/135050689500200305.

Adams, Vincanne, Michelle Murphy, and Adele E. Clarke. 2009. "Anticipation: Technoscience, Life, Affect, Temporality." *Subjectivity* 28 (1): 246–65. https://doi.org/10.1057/sub.2009.18.

Ahmed, Sara. 2020. *Queer Phenomenology: Orientations, Objects, Others*. Durham: Duke University Press.

Akrich, Madeleine. 1992. "The De-Scription of Technical Objects." In *Shaping Technology/Building Society: Studies in Societechnical Change*, edited by Wiebe E. Bijker and John Law, 205–24. Cambridge and London: MIT Press.

Akyüz, Kaya, Gauthier Chassang, Melanie Goisauf, Łukasz Kozera, Signe Mezinska, Olga Tzortzatou, and Michaela Th Mayrhofer. 2021. "Biobanking and Risk Assessment: A Comprehensive Typology of Risks for an Adaptive Risk Governance." *Life Sciences, Society and Policy* 17 (10): 1–28. https://doi.org/10.1186/s40504-021-00117-7.

Akyüz, Kaya, Mónica Cano Abadía, Melanie Goisauf, and Michaela Th. Mayrhofer. 2024. "Unlocking the Potential of Big Data and AI in Medicine: Insights from Biobanking." *Frontiers in Medicine* 11. https://doi.org/10.3389/fmed.2024.1336588.

American Medical Association. 2018. *AMA Adopts New Policies at 2018 Interim Meeting*. National Harbor, MD.

Blackless, Melanie, Anthony Charuvastra, Amanda Derryck, Anne Fausto-Sterling, Karl Lauzanne, and Ellen Lee. 2000. "How Sexually Dimorphic Are We? Review and Synthesis." *American Journal of Human Biology* 12 (2): 151–66. https://doi.org/10.1002/(SICI)1520-6300(200003/04)12:2<151::AID-AJHB1>3.0.CO;2-F.

Bowker, Geoffrey C. 2005. *Memory Practices in the Sciences*. Cambridge and London: MIT Press.
Bowker, Geoffrey C., and Susan Leigh Star. 2000. *Sorting Things Out: Classification and Its Consequences*. Edited by Wiebe E. Bijker, W. Bernard Carlson and Trevor Pinch, Inside Technology. Cambridge and London: MIT Press.
Brady, Adrian P., and Emanuele Neri. 2020. "Artificial Intelligence in Radiology – Ethical Considerations." *Diagnostics* 10 (4). https://doi.org/10.3390/diagnostics10040231.
Brown, Nik. 2003. "Hope Against Hype-Accountability in Biopasts, Presents and Futures." *Science & Technology Studies* 28 (2). https://doi.org/10.23987/sts.55152.
Butler, Judith. 1990. *Gender Trouble. Feminism and the Subversion of Identity*. New York and London: Routledge.
Butler, Judith. 1996. *Bodies That Matter: On the Discursive Limits of Sex*. New York and London: Routledge.
Cano Abadía, Mónica. 2021. *Judith Butler: Performatividad y Vulnerabilidad*. Barcelona: Shackleton Books.
Chaudhary, V. Bala, and Asmeret Asefaw Berhe. 2020. "Ten Simple Rules for Building an Antiracist Lab." *PLoS Computational Biology* 16 (10): e1008210. https://doi.org/10.1371/journal.pcbi.1008210.
Chen, Richard J., Ming Y. Lu, Tiffany Y. Chen, Drew F. K. Williamson, and Faisal Mahmood. 2021. "Synthetic Data in Machine Learning for Medicine and Healthcare." *Nature Biomedical Engineering* 5 (6): 493–7. https://doi.org/10.1038/s41551-021-00751-8.
Cirillo, Davide, Silvina Catuara-Solarz, Czuee Morey, Emre Guney, Laia Subirats, Simona Mellino, Annalisa Gigante, Alfonso Valencia, María José Rementeria, Antonella Santuccione Chadha, and Nikolaos Mavridis. 2020. "Sex and Gender Differences and Biases in Artificial Intelligence for Biomedicine and Healthcare." *NPJ Digital Medicine* 3 (1): 81. https://doi.org/10.1038/s41746-020-0288-5.
Clarke, Adele E., Melanie Jeske, Laura Mamo, and Janet K. Shim. 2021. "Biomedicalization Revisited." In *The Wiley Blackwell Companion to Medical Sociology*, 125–49. Oxford: Wiley Blackwell.
Clarke, Adele E., Laura Mamo, Jennifer R. Fishman, Janet K. Shim, and Jennifer Ruth Fosket. 2003. "Biomedicalization: Technoscientific Transformations of Health, Illness, and U.S. Biomedicine." *American Sociological Review* 68 (2): 161–94. https://doi.org/10.2307/1519765.
Coeckelbergh, Mark. 2020. *AI Ethics*. Cambridge, MA: MIT Press.
Collins, Francis S., and Harold Varmus. 2015. "A New Initiative on Precision Medicine." *New England Journal of Medicine* 372 (9): 793–5. https://doi.org/10.1056/NEJMp1500523.
Corrigan, Oonagh P. 2002. "'First in Man': The Politics and Ethics of Women in Clinical Drug Trials." *Feminist Review* 72 (1): 40–52. https://doi.org/10.1057/palgrave.fr.9400055.
Couldry, Nick, and Ulises Mejias. 2018. "Data Colonialism: Rethinking Big Data's Relation to the Contemporary Subject." *Television & New Media* 20 (4): 336–49. https://doi.org/10.1177/1527476418796632.
Davenport, Thomas, and Ravi Kalakota. 2019. "The Potential for Artificial Intelligence in Healthcare." *Future Healthcare Journal* 6 (2): 94. https://doi.org/10.7861/futurehosp.6-2-94.
Delgado, Janet, Mónica Cano Abadía, Kaya Akyüz, Melanie Goisauf, and David Rodríguez-Arias. 2024. "Uncovering the Persistent Gap: The Ongoing Challenge of Integrating Sex and Gender in Biomedical Research." *The Journal of Biomedical Research*. https://doi.org/10.7555/JBR.38.20240157.
Demir, Ipek, and Madeleine J. Murtagh. 2013. "Data Sharing Across Biobanks: Epistemic Values, Data Mutability and Data Incommensurability." *New Genetics and Society* 32 (4): 350–65. https://doi.org/10.1080/14636778.2013.846582.
D'Ignazio, Catherine, and Lauren F. Klein. 2020. *Data Feminism*. Cambridge, MA: MIT Press.

DuBois, L. Zachary, and Heather Shattuck-Heidorn. 2021. "Challenging the Binary: Gender/Sex and the Bio-Logics of Normalcy." *American Journal of Human Biology* 33 (5): e23623. https://doi.org/10.1002/ajhb.23623.

Epstein, Steven. 2007. *Inclusion: The Politics of Difference in Medical Research*. Chicago and London: The University of Chicago Press.

European Commission. 2021. *Ethics By Design and Ethics of Use Approaches for Artificial Intelligence*. https://ec.europa.eu/info/funding-tenders/opportunities/docs/2021-2027/horizon/guidance/ethics-by-design-and-ethics-of-use-approaches-for-artificial-intelligence_he_en.pdf (24.07.2023).

Fausto-Sterling, Anne. 2000. *Sexing the Body: Gender Politics and the Construction of Sexuality*. New York: Basic Books.

Fausto-Sterling, Anne. 2012. *Sex/Gender: Biology in a Social World, the Routledge Series Integrating Science and Culture*. New York: Routledge.

Fausto-Sterling, Anne. 2018. "Why Sex Is Not Binary." *The New York Times*. https://www.nytimes.com/2018/10/25/opinion/sex-biology-binary.html (10.03.2022).

Fleck, Ludwik. 1981 [1935]. *Genesis and Development of a Scientific Fact*. Chicago and London: University of Chicago Press.

Floridi, Luciano. 2020. "What the Near Future of Artificial Intelligence Could Be." In *The 2019 Yearbook of the Digital Ethics Lab*, edited by Christopher Burr and Silvia Milano, 127–42. Cham: Springer International Publishing.

Foucault, Michel. 1972. *Archaeology of Knowledge and the Discourse on Language*. New York: Pantheon Books.

Foucault, Michel. 1977. "Nietzsche, Genealogy, History." In *Language, Counter-Memory, Practice: Selected Essays and Interviews*, edited by D. F. Bouchard, 139–64. Ithaca: Cornell University Press.

Foucault, Michel. 1980. "Truth and Power." In *Power/Knowledge. Selected Interviews and Other Writings. 1972–1977*, edited by Colin Gordon, 109–32. New York: Pantheon Books.

Fröhlich, Holger, Rudi Balling, Niko Beerenwinkel, Oliver Kohlbacher, Santosh Kumar, Thomas Lengauer, Marloes H. Maathuis, Yves Moreau, Susan A. Murphy, Teresa M. Przytycka, Michael Rebhan, Hannes Röst, Andreas Schuppert, Matthias Schwab, Rainer Spang, Daniel Stekhoven, Jimeng Sun, Andreas Weber, Daniel Ziemek, and Blaz Zupan. 2018. "From Hype to Reality: Data Science Enabling Personalized Medicine." *BMC Medicine* 16 (1): 150. https://doi.org/10.1186/s12916-018-1122-7.

Ganna, Andrea, Karin J. H. Verweij, Michel G. Nivard, Robert Maier, Robbee Wedow, Alexander S. Busch, Abdel Abdellaoui, Shengru Guo, J. Fah Sathirapongsasuti, and 23andMe Research Team. 2019. "Large-scale GWAS Reveals Insights into the Genetic Architecture of Same-Sex Sexual Behavior." *Science* 365 (6456): eaat7693. https://doi.org/10.1126/science.aat7693.

Geels, Frank W., and Wim A. Smit. 2000. "Failed Technology Futures: Pitfalls and Lessons from a Historical Survey." *Futures* 32 (9): 867–85. https://doi.org/10.1016/S0016-3287(00)00036-7.

Geis, J. Raymond, Adrian P. Brady, Carol C. Wu, Jack Spencer, Erik Ranschaert, Jacob L. Jaremko, Steve G. Langer, Andrea Borondy Kitts, Judy Birch, William F. Shields, Robert van den Hoven van Genderen, Elmar Kotter, Judy Wawira Gichoya, Tessa S. Cook, Matthew B. Morgan, An Tang, Nabile M. Safdar, and Marc Kohli. 2019. "Ethics of Artificial Intelligence in Radiology: Summary of the Joint European and North American Multisociety Statement." *Canadian Association of Radiologists Journal* 70 (4): 329–34. https://doi.org/10.1016/j.carj.2019.08.010.

Geneviève, Lester Darryl, Andrea Martani, David Shaw, Bernice Simone Elger, and Tenzin Wangmo. 2020. "Structural Racism in Precision Medicine: Leaving No One Behind." *BMC Medical Ethics* 21 (1): 17. https://doi.org/10.1186/s12910-020-0457-8.

Ghassemi, Marzyeh, and Elaine Okanyene Nsoesie. 2022. "In Medicine, How Do We Machine Learn Anything Real?" *Patterns* 3 (1): 100392. https://doi.org/10.1016/j.patter.2021.100392.

Gitelman, Lisa, ed. 2013. *"Raw Data" Is an Oxymoron*. Edited by Geoffrey C. Bowker and Paul N. Edwards, Infrastructures Series. Cambridge and London: MIT Press.

Goisauf, Melanie, and Mónica Cano Abadía. 2022. "Ethics of AI in Radiology: A Review of Ethical and Societal Implications." *Frontiers in Big Data* 5 (850383). https://doi.org/10.3389/fdata.2022.850383.

Goisauf, Melanie, Kaya Akyüz, and Gillian M. Martin. 2020. "Moving Back to the Future of Big Data-Driven Research: Reflecting on the Social in Genomics." *Humanities and Social Sciences Communications* 7 (1): 55. https://doi.org/10.1057/s41599-020-00544-5.

Haraway, Donna J. 1988. "Situated Knowledges: The Science Question in Feminism and the Privilege of Partial Perspective." *Feminist Studies* 14 (3): 575–99. https://doi.org/10.2307/3178066.

Hoeyer, Klaus. 2019. "Data as Promise: Reconfiguring Danish Public Health Through Personalized Medicine." *Social Studies of Science* 49 (4): 531–55. https://doi.org/10.1177/0306312719858697.

Holub, Petr, Florian Kohlmayer, Fabian Prasser, Michaela Th Mayrhofer, Irene Schlünder, Gillian M. Martin, Sara Casati, Lefteris Koumakis, Andrea Wutte, Łukasz Kozera, Dominik Strapagiel, Gabriele Anton, Gianluigi Zanetti, Osman Ugur Sezerman, Maimuna Mendy, Dalibor Valík, Marialuisa Lavitrano, Georges Dagher, Kurt Zatloukal, GertJan B. van Ommen, and Jan-Eric Litton. 2018. "Enhancing Reuse of Data and Biological Material in Medical Research: From FAIR to FAIR-Health." *Biopreservation and Biobanking* 16 (2): 97–105. https://doi.org/10.1089/bio.2017.0110.

ISO. 2012. *15189:2012 Medical Laboratories – Requirements for Quality and Competence*. Geneva: International Organization for Standardization.

ISO. 2018. *20387: 2018 Biotechnology, Biobanking, General Requirements for Biobanking*. Geneva: International Organization for Standardization.

ISO, and IEC. 2013. *ISO/IEC 27001: 2013 Information Technology – Security Techniques – Information Security Management Systems – Requirements*. Geneva: International Organization for Standardization.

Jasanoff, Sheila. 2004. "The Idiom of Co-Production." In *States of Knowledge: The Co-Production of Science and Social Order*, edited by S. Jasanoff, 1–12. London and New York: Routledge.

Jones, Nat C., Amy K. Otto, Dana E. Ketcher, Jennifer B. Permuth, Gwendolyn P. Quinn, and Matthew B. Schabath. 2020. "Inclusion of Transgender and Gender Diverse Health Data in Cancer Biorepositories." *Contemporary Clinical Trials Communications* 19: 100597. https://doi.org/10.1016/j.conctc.2020.100597.

Jones, Nat C., Monica E. Reyes, Gwendolyn P. Quinn, and Matthew B. Schabath. 2020. "Survey of Principal Investigators in Biobanking: Knowledge, Attitudes, and Research Behaviors About Transgender and Gender-Diverse Patients." *JCO Oncology Practice* 16 (10): e1192–e1201. https://doi.org/10.1200/OP.20.00193.

Kacel, Elizabeth L., Tyson L. Pankey, Jennifer A. Vencill, Dagoberto Heredia, Cesar A. Gonzalez, Kathryn J. Ruddy, and Shawna L. Ehlers. 2022. "Include, Affirm, and Empower: A Paradigm Shift in Cancer Clinical Trials for Sexual and Gender Diverse Populations." *Annals of LGBTQ Public and Population Health* 3 (1):18–40. https://doi.org/10.1891/LGBTQ-2021-0013.

Karasti, Helena, and Jeanette Blomberg. 2018. "Studying Infrastructuring Ethnographically." *Computer Supported Cooperative Work (CSCW)* 27 (2): 233–65. https://doi.org/10.1007/s10606-017-9296-7.

Kaye, Jane. 2011. "From Single Biobanks to International Networks: Developing E-Governance." *Human Genetics* 130: 377–82. https://doi.org/10.1007/s00439-011-1063-0.

Keyes, Os, Zoë Hitzig, and Mwenza Blell. 2021. "Truth from the Machine: Artificial Intelligence and the Materialization of Identity." *Interdisciplinary Science Reviews* 46 (1–2): 158–75. https://doi.org/10.1080/03080188.2020.1840224.

Kim, Alison M., Candace M. Tingen, and Teresa K. Woodruff. 2010. "Sex Bias in Trials and Treatment Must End." *Nature* 465 (7299): 688–9. https://doi.org/10.1038/465688a.

Kitchin, Rob. 2014. *The Data Revolution: Big Data, Open Data, Data Infrastructures & Their Consequences*. London and Thousand Oaks: SAGE.

Knorr-Cetina, Karin. 1981. *The Manufacture of Knowledge: An Essay on the Constructivist and Contextual Nature of Science*. Oxford: Pergamon Press.

Laqueur, Thomas. 1992. *Making Sex. Body and Gender from the Greeks to Freud*. Cambridge, MA and London: Harvard University Press.

Larrazabal, Agostina J., Nicolás Nieto, Victoria Peterson, Diego H. Milone, and Enzo Ferrante. 2020. "Gender Imbalance in Medical Imaging Datasets Produces Biased Classifiers for Computer-Aided Diagnosis." *Proceedings of the National Academy of Sciences* 117 (23): 12592–4. https://doi.org/10.1073/pnas.1919012117.

Leonelli, Sabina. 2020. "Learning from Data Journeys." In *Data Journeys in the Sciences*, edited by Sabina Leonelli and Niccolò Tempini, 1–24. Cham: Springer International Publishing.

Macklin, Ruth. 2010. "Enrolling Pregnant Women in Biomedical Research." *The Lancet* 375 (9715): 632–3. https://doi.org/10.1016/S0140-6736(10)60257-7.

Martin, Emily. 2001. *The Woman in the Body: A Cultural Analysis of Reproduction*. Boston: Beacon Press.

Maas, Samantha A., Nicholas J. Wiesenthal, Sara E. Brownell, and Katelyn M. Cooper. 2024. "Disrupting the Master Narrative in Academic Biology as LGBTQ+ Ph.D. Students: Learning, Teaching, and Conducting Research." *CBE—Life Sciences Education* 23 (4). https://doi.org/10.1187/cbe.24-02-0058.

Mayrhofer, Michaela Th. 2013. "About the New Significance and the Contingent Meaning of Biological Material and Data in Biobanks." *History and Philosophy of the Life Sciences* 35 (3): 449–67.

McKinney, Scott Mayer, Marcin Sieniek, Varun Godbole, Jonathan Godwin, Natasha Antropova, Hutan Ashrafian, Trevor Back, Mary Chesus, Greg C. Corrado, Ara Darzi, Mozziyar Etemadi, Florencia Garcia-Vicente, Fiona J. Gilbert, Mark Halling-Brown, Demis Hassabis, Sunny Jansen, Alan Karthikesalingam, Christopher J. Kelly, Dominic King, Joseph R. Ledsam, David Melnick, Hormuz Mostofi, Lily Peng, Joshua Jay Reicher, Bernardino Romera-Paredes, Richard Sidebottom, Mustafa Suleyman, Daniel Tse, Kenneth C. Young, Jeffrey De Fauw, and Shravya Shetty. 2020. "International Evaluation of an AI System for Breast Cancer Screening." *Nature* 577 (7788): 89–94. https://doi.org/10.1038/s41586-019-1799-6.

McQuillan, Dan. 2022. *Resisting AI: An Anti-Fascist Approach to Artificial Intelligence*. Bristol: Bristol University Press.

Meijer, Irene Costera, and Baukje Prins. 1998. "How Bodies Come to Matter: An Interview with Judith Butler." *Signs: Journal of Women in Culture and Society* 23 (2): 275–86. https://doi.org/10.1086/495251.

Merino-Martinez, Roxana, Loreana Norlin, David van Enckevort, Gabriele Anton, Simone Schuffenhauer, Kaisa Silander, Linda Mook, Petr Holub, Raffael Bild, Morris Swertz, and Jan-Eric Litton. 2016. "Toward Global Biobank Integration by Implementation of the Minimum Information About BIobank Data Sharing (MIABIS 2.0 Core)." *Biopreservation and Biobanking* 14 (4): 298–306. https://doi.org/10.1089/bio.2015.0070.

Montañez, Amanda. 2017. "Beyond XX and XY." *Scientific American* 317 (3): 50–1. https://doi.org/10.1038/scientificamerican0917-50.

Murtagh, M. J., M. Machirori, C. L. Gaff, M. T. Blell, J. de Vries, M. Doerr, E. S. Dove, A. Duncanson, J. Hastings Ward, R. Hendricks-Sturrup, C. W. L. Ho, A. Johns, Y. Joly, K. Kato, K. Katsui, J. Kumuthini, F. Maleady-Crowe, A. Middleton, R. Milne, J. T. Minion, M. Matshaba, S. Mulrine, C. Patch, R. Ryan, and W. Viney. 2021. "Engaged Genomic Science Produces Better and Fairer Outcomes: An Engagement Framework for Engaging and Involving Participants, Patients and Publics in Genomics Research and Healthcare Implementation." *Wellcome Open Research* 6: 311. https://doi.org/10.12688/wellcomeopenres.17233.1.

Nietzsche, Friedrich. 1997 [1873]. "On Truth and Falsity in Their Extramoral Sense." In *Philosophical Writings: Friedrich Nietzsche*, edited by Reinhold Grimm and Caroline Molina y Vedia, 87–99. New York: Continuum.

Noble, Safiya Umoja. 2018. *Algorithms of Oppression. How Search Engines Reinforce Racism*. New York: New York University Press.
Obermeyer, Ziad, Brian Powers, Christine Vogeli, and Sendhil Mullainathan. 2019. "Dissecting Racial Bias in an Algorithm Used to Manage the Health of Populations." *Science* 366 (6464): 447–53. https://doi.org/10.1126/science.aax2342.
Posey, Jennifer E., Anne H. O'Donnell-Luria, Jessica X. Chong, Tamar Harel, Shalini N. Jhangiani, Zeynep H. Coban Akdemir, Steven Buyske, Davut Pehlivan, Claudia M. B. Carvalho, Samantha Baxter, Nara Sobreira, Pengfei Liu, Nan Wu, Jill A. Rosenfeld, Sushant Kumar, Dimitri Avramopoulos, Janson J. White, Kimberly F. Doheny, P. Dane Witmer, Corinne Boehm, V. Reid Sutton, Donna M. Muzny, Eric Boerwinkle, Murat Günel, Deborah A. Nickerson, Shrikant Mane, Daniel G. MacArthur, Richard A. Gibbs, Ada Hamosh, Richard P. Lifton, Tara C. Matise, Heidi L. Rehm, Mark Gerstein, Michael J. Bamshad, David Valle, James R. Lupski, and Genomics Centers for Mendelian. 2019. "Insights into Genetics, Human Biology and Disease Gleaned from Family Based Genomic Studies." *Genetics in Medicine* 21 (4): 798–812. https://doi.org/10.1038/s41436-018-0408-7.
Prainsack, Barbara. 2015. "Is Personalized Medicine Different? (Reinscription: The Sequel) A Response to Troy Duster." *The British Journal of Sociology* 66 (1): 28–35. https://doi.org/10.1111/1468-4446.12117.
Quinn, Thomas P., Stephan Jacobs, Manisha Senadeera, Vuong Le, and Simon Coghlan. 2022. "The Three Ghosts of Medical AI: Can the Black-Box Present Deliver?" *Artificial Intelligence in Medicine* 124: 102158. https://doi.org/10.1016/j.artmed.2021.102158.
Rasheed, Khansa, Adnan Qayyum, Mohammed Ghaly, Ala Al-Fuqaha, Adeel Razi, and Junaid Qadir. 2022. "Explainable, Trustworthy, and Ethical Machine Learning for Healthcare: A Survey." *Computers in Biology and Medicine* 149: 106043. https://doi.org/10.1016/j.compbiomed.2022.106043.
Rettberg, Jill Walker. 2020. "Situated Data Analysis: A New Method for Analysing Encoded Power Relationships in Social Media Platforms and Apps." *Humanities and Social Sciences Communications* 7 (1): 5. https://doi.org/10.1057/s41599-020-0495-3.
Ritz, Stacey A., and Lorraine Greaves. 2022. "Transcending the Male-Female Binary in Biomedical Research: Constellations, Heterogeneity, and Mechanism When Considering Sex and Gender." *International Journal of Environmental Research and Public Health* 19 (7). https://doi.org/10.3390/ijerph19074083.
Rose, Nikolas. 2007. *The Politics of Life Itself: Biomedicine, Power, and Subjectivity in the Twenty-First Century*. Princeton and Oxford: Princeton University Press.
Rubin, Gayle. 1975. "The Traffic in Women: Notes on the 'Political Economy' of Sex." In *Toward an Anthropology of Women*, edited by Rayna R. Reiter, 157–210. New York: Monthly Review Press.
Schiebinger, Londa, and Ineke Klinge. 2020. *Gendered Innovations 2: How Inclusive Analysis Contributes to Research and Innovation*. Luxembourg: Publications Office of the European Union.
Sedgwick, Eve Kosofsky. 2008. *Epistemology of the Closet*. Berkeley: University of California Press.
Tam, Vivian, Nikunj Patel, Michelle Turcotte, Yohan Bossé, Guillaume Paré, and David Meyre. 2019. "Benefits and Limitations of Genome-Wide Association Studies." *Nature Reviews Genetics* 20 (8): 467–84. https://doi.org/10.1038/s41576-019-0127-1.
UNESCO. 2020. *Artificial Intelligence and Gender Equality: Key Findings of UNESCO's Global Dialogue*. Paris: United Nations Educational, Scientific and Cultural Organization.
Warner, Michael. 1991. "Introduction: Fear of a Queer Planet." *Social Text* 29: 3–17.
Wilkinson, Mark D., Michel Dumontier, IJsbrand Jan Aalbersberg, Gabrielle Appleton, Myles Axton, Arie Baak, Niklas Blomberg, Jan-Willem Boiten, Luiz Bonino da Silva Santos, Philip E. Bourne, Jildau Bouwman, Anthony J. Brookes, Tim Clark, Mercè Crosas, Ingrid Dillo, Olivier Dumon, Scott Edmunds, Chris T. Evelo, Richard Finkers, Alejandra Gonzalez-Beltran, Alasdair J. G. Gray, Paul Groth, Carole Goble, Jeffrey S. Grethe, Jaap Heringa, Peter A. C. 't Hoen, Rob Hooft, Tobias Kuhn, Ruben Kok, Joost

Kok, Scott J. Lusher, Maryann E. Martone, Albert Mons, Abel L. Packer, Bengt Persson, Philippe Rocca-Serra, Marco Roos, Rene van Schaik, Susanna-Assunta Sansone, Erik Schultes, Thierry Sengstag, Ted Slater, George Strawn, Morris A. Swertz, Mark Thompson, Johan van der Lei, Erik van Mulligen, Jan Velterop, Andra Waagmeester, Peter Wittenburg, Katherine Wolstencroft, Jun Zhao, and Barend Mons. 2016. "The FAIR Guiding Principles for Scientific Data Management and Stewardship." *Scientific Data* 3 (1): 160018. https://doi.org/10.1038/sdata.2016.18.

Zook, Matthew, Solon Barocas, Danah Boyd, Kate Crawford, Emily Keller, Seeta Peña Gangadharan, Alyssa Goodman, Rachelle Hollander, Barbara A. Koenig, Jacob Metcalf, Arvind Narayanan, Alondra Nelson, and Frank Pasquale. 2017. "Ten Simple Rules for Responsible Big Data Research." *PLOS Computational Biology* 13 (3): e1005399. https://doi.org/10.1371/journal.pcbi.1005399.

Part 3

Communicating the Unknown

6 Fairness of an AI System in the Case of a Biobank of Images and Imaging Biomarkers

Valentina Colcelli

1 Introduction

This chapter approaches the specific case of the application of artificial intelligence (AI) in imaging biobanks for research activities because of the particular use made of AI in imaging biobanks. Many ethical and legal issues that focus on the balance between research activity and the rights and freedoms of individuals are related to research biobanks. The creation and management of biobank collections give rise to several ethical and juridical issues and may provoke societal concerns (Gaskell et al. 2000). Research biobanks are increasingly common. The development of biobanks facilitates scientific research with biological samples or images and their associated personal data. However, there is currently a lack of systematic rules to address the issues surrounding the creation of a biobank, and specifically of an imaging biobank. The increasing number of research biobanks requires a specific legal framework for biobanks at the European Union (EU) level – or a dedicated soft law (e.g., a Code of Ethics) – to address the key issues that arise in the daily life of a biobanking activity. These issues include:

(i) The definition of the type of consent the donor/patient/research participant must give (e.g., broad consent or dynamic consent as a common standard) because of "future research" as a structural question for research biobanks.
(ii) Rules addressing data protection.
(iii) A common framework to facilitate common answers from different research ethics committees.
(iv) Some provisions linked to the standardization of quality and safety.

The 2012 EU Commission Report of the Expert Group on Dealing With Ethical and Regulatory Challenges of International Biobank Research has underlined the necessity of harmonizing the rules and the legal and ethical practices which govern the management of research in the European Union. European Union law provides a very developed regulatory framework for biomedical activities; examples are Regulation (EU) 536/2014 on Clinical Trials on Medicinal Products for Human Use, and Regulation EU 2016/679 on the protection of personal data (EU General Data Protection Regulation or GDPR). However, these legal sources, like other

DOI: 10.4324/9781032634487-10

EU laws, do not specifically take the research activities of biobanks into consideration and instead leave regulation to the Member States. The increasing number of research biobanks and the importance of their role in supporting medical and biological research have resulted in actions aimed at the development and sharing of best practices and benchmarking standards for biobanking which guarantee the fundamental rights of the donors/patients/research participants and increase trust in the system while not placing unnecessary obstacles in the way of biomedical research.

At the same time, only a few of the EU Member States' legal systems have adopted a special law on research biobanks. In many legal systems, which are sometimes characterized as hybrid systems, in the spaces left empty by the law, the regulation of the establishment of biobanks, the organizational needs of the infrastructure, and the management of biological samples or images, as well as the protection of the rights of stakeholders, which have to be reconciled with the needs of science, are often the result of interpretations of individuals and soft law instruments that operate according to principles and are often formed by practitioners. In the absence of a specific and common regulatory framework, the question arises as to how the existing legislation from other sectors should be properly used (Zika et al. 2010). In particular, we appeal to data protection rules (Staunton et al. 2019; Slokenberga 2021; Mitrou 2018) and clinical trial provisions contained in national civil codes. Although these schemes have enabled practitioners to map out a legal horizon within which to move, they do not always provide an adequate regime. Each borrowed model may be deficient or inappropriate, albeit for different reasons.

At this stage of knowledge about all types of research biobanks, we must also take into consideration the impact of artificial intelligence,[1] with a special focus on imaging biobanks. In the case of imaging biobanks, it is very well documented in research studies (Seyyed-Kalantari et al. 2021; Park et al. 2020; Hacker 2018; Geis et al. 2019) that if an AI algorithm is trained with imaging data that are unbalanced with respect to sex, socioeconomic characteristics, or ethnicity, and the algorithm is given the health differences within and across these groups, the resulting model is likely to lead to biased predictions. If the problem of bias in AI is common to all medical applications, it is even more problematic in medical imaging because personal attributes such as sex, age, ethnicity, and socioeconomic characteristics are not always retained during data preparation in order to minimize the possibility of patient identification (Diaz et al. 2021). On the other hand, despite the lack of retention of personal data, the image may still allow the sex, age, or ethnicity of the person to be traced even when this is not necessary for the purpose for which the system is being trained. This also has an impact on the process of collecting and preparing datasets, because the process should be transparent and properly documented through informed consent during participant recruitment.

The ethical and legal issues related to an innovative AI technique include socio-legal challenges regarding the requirements that should be met by any AI system if it is to be fair and trustworthy. The juridical and ethical component to be addressed is not limited to the treatment of personal data but, more importantly, extends to the intelligibility and transparency of the reasoning process. The main idea is that AI algorithms should incorporate the principles of democracy, the rule of law, and the fundamental rights arising from the EU legal system, from the design stage onward.

2 Imaging Biobanks

A collection of biological material which usually and originally consists of tissues is at the core of the "infrastructure" for scientific research which qualifies as a biobank (see the OECD (2007) notion of a biological resources center and the European Union law idea of research infrastructure[2]). A biobank may be defined as "any collection of biological materials, whether the source be human, plant, or animal, fungi, bacteria, microorganisms or other living families, as well as bioinformatic data on such organic materials" (Perry 2012, 267) or as

> a non-profit service unit for the collection, processing, storage and distribution of human biological samples and related data for research and diagnosis, . . . officially recognized by the competent health authorities, [that] applies a quality system and guarantees the rights of those involved.[3]
>
> (BBMRI.it n.d.)

In recent years, imaging data have been included within the notion of biobanks; thus, it is possible to talk about imaging biobanks. According to the definition established by the Imaging Biobanks Working Group of the European Society of Radiology in 2014 (European Society of Radiology 2015), we can describe imaging biobanks as "organized databases of medical images and associated imaging biomarkers, shared among multiple researchers, linked to other biorepositories". Starting from this definition, an imaging biobank is not just an image archiving and transmission system like the PACS (picture archiving and communication system) which is used in hospitals, nor is it just a platform that allows the storage and retrieval of medical images and associated metadata. In an imaging biobank, biomarkers are extracted, for instance, through radiomic analysis of imaging data (typically radiological imaging data), in contrast to clinical data (molecular, biological, and genomic data) processed in a typical tissue biobank.

The great potential of imaging biobanks is the ability to provide multi-omic data for the study of advanced imaging techniques on imaging pools with a sufficient sample size. This availability of data is necessary for researchers to find an association between phenotype and genotype, and to design and validate new imaging biomarkers, as well as to understand their biological significance, which can be a crucial point in precision medicine. Once large, high-quality, and well-curated datasets are available within a biobank, they can be used for data mining, which refers to the process of discovering patterns in large datasets that can lead to new biomarkers (Borgheresi et al. 2022, 53).

AI systems in such biobanks are mainly used to transform the old qualitative or semiquantitative clinical imaging into fully used data-driven, quantitative, prospective thinking in radiomics.

The use of AI for understanding medical imaging (especially in radiology) is still quite baffling in several respects because AI allows:

(i) Data acquisition and preprocessing (uniform protocols across clinical sites, and dependence on the device and operators).
(ii) Data quality (curation, annotation, and segmentation).
(iii) Data protection for both image data/metadata and associated health records.
(iv) The management of known and unknown biases.
(v) Continuous monitoring of model/concept/performance drift.
(vi) The reproducibility, accountability, explainability, and interpretability of model predictions.

In other words, even if strategies exist for balancing and augmenting the dataset, or for detecting and managing biases, not all of these strategies could be applied in the medical/clinical setting; also, well-known phenomena, such as model overfitting and the underestimation of the assumptions of statistical hypotheses (rarely verified), could occur and badly affect the knowledge extracted from radiological images (Borgheresi et al. 2022).

As a matter of fact, three approaches are used to design radiation therapy treatment plans: the classic approach, the so-called modern approach, and an AI-based approach. In the first of these, medical physicists and radiotherapists design the treatment plan using their medical experience, in the second, a physics-based algorithm designs the treatment plan using equations and the plan is checked by medical physicists and radiotherapists, and then in the last approach an AI-based algorithm designs the treatment plan, which is checked by medical physicists and radiotherapists. The treatment plan is checked by performing experimental measurements on phantoms or in vivo (Barucci 2022). Imaging biobanks offer data and biomarkers to develop and train machine learning or networking learning systems to:

- "Classify: characterization of tissue/disease/phenotype.
- Segment: lesion/tissues/organs.
- Achieve other things, such as super-resolution of image, data generation, and simulations".

(Barucci 2022)

In this context and because of the treatment of clinical imaging by AI systems, more and more software for diagnosis/therapy/treatment will be based on such technology. Thus, this chapter is a reflection on the legal and ethical aspects that should be addressed in the creation of imaging biobanks and the application of AI in developing imaging biomarkers.

3 Methodology

To achieve the purposes set out in Section 1, the chapter will use the report titled "FUTURE-AI: Guiding Principles and Consensus Recommendations for Trustworthy Artificial Intelligence in Medical Imaging" (Lekadir et al. 2021) as a starting point to try to address the main issues related to fairness in the application of AI systems in imaging biobanks in the EU legal system. The report reproduces the principles and the related recommendations and best practices that were defined by building on the experiences and results accumulated from five large European projects on AI in health imaging (the AI4HI Network, comprising the EuCanImage, PRIMAGE, CHAIMELEON, INCISIVE, and ProCAncer-I projects). Given the lack of a specific regulation for the biobank phenomenon, the experience of others becomes precious. In this framework, the experience related to important research projects is analyzed and taken into consideration.

These five large European projects on AI in health imaging set out guiding principles named FUTURE-AI (Lekadir et al. 2021) to address the problems of (a) fairness, (b) universality, (c) traceability, (d) usability, (e) robustness, and (f) explainability. Those principles were further translated into a framework of concrete recommendations for specifying, developing, evaluating, and deploying into clinical practice AI solutions that are technically, clinically, and ethically trustworthy. This chapter tries to approach these guiding principles from a legal point of view.

This chapter analyzes the substantive and procedural dimension of fairness as referred to in the European Commission's "Ethics Guidelines for Trustworthy AI" (2019) (hereafter known as the Ethics Guidelines) and integrated with the definition in the FUTURE-AI guidelines on fairness (Section 4). Because of the many different interpretations of fairness, we apply the definition of fairness from the Ethics Guidelines, as this provides a common EU framework for addressing the concept of fairness arising from the FUTURE-AI guidelines. After a brief reminder of the relationship between the GDPR and imaging in the context of imaging biobanks and imaging biomarkers (Section 5), the issue of labelling metadata correctly to achieve the appropriate representation of patient groups in the training of imaging datasets, which arises because of both the pseudoanonymization and the anonymization of medical images, will be addressed (Section 6).

Next, the issue of training AI imaging solutions under European non-discrimination law will be addressed (Section 7), because if an AI algorithm is trained with imaging data that are unbalanced with respect to gender, socioeconomic characteristics (Glover et al. 2017),[4] or ethnicity, and given the health differences within and between these groups, the resulting model is likely to lead to biased predictions. In this case, to avoid problems, it is suggested – among other things – that the evidentiary requirements used by the Court of Justice of the European Union be applied to the specific case of AI in biobanks of images.

Finally, it will be emphasized how fairness can be compromised by overreliance on AI systems by both experienced and less experienced radiologists (Povyakalo et al. 2013).

4 Analyzing the Fairness of Health Imaging Under the Guiding Principles of FUTURE-AI and the "Ethics Guidelines for Trustworthy AI"

The first principle addressed by the FUTURE-AI guidelines is fairness/equity. Focusing on fairness and its legally relevant aspects is very useful, not least because the elements derived from the analysis of fairness have a direct impact on the other principles (universality, traceability, usability, robustness, and explicability). With specific attention to the trustworthiness of AI in medicine and equitable AI in medical imaging, the FUTURE-AI guidelines define fairness as follows:

> medical AI algorithms should maintain the same performance when applied to similarly situated individuals (individual fairness) and across subgroups of individuals, including under-represented groups (group fairness). Healthcare, which is an expensive but critical service for society, should be provided equally for all patients independently of their gender, ethnicity, income and geography. AI algorithms should not exacerbate existing health disparities, but instead should facilitate and enhance access to high-quality radiology services for all individuals and groups. Medical AI algorithms should be built such that they address common as well as hidden biases in training datasets.
> (Lekadir et al. 2021)

Starting from this definition, we will approach the main issues arising from it and the Ethics Guidelines (European Commission 2019) to give us a common framework under EU soft law. The Ethics Guidelines approach fairness by looking at two dimensions: the substantive and the procedural.

The substantive dimension of fairness is referred to in the Ethics Guidelines and is integrated into the definition of fairness in the FUTURE-AI guidelines. It implies a commitment to "ensuring equal and just distribution of both benefits and costs" (European Commission 2019) of medical services that use AI systems, as well as guaranteeing "that individuals and groups are free from unfair bias, discrimination and stigmatisation" (European Commission 2019) when they access healthcare services that apply AI to health imaging. Discrimination and stigmatization must be avoided if they result from the use of AI systems.

In addition, those who use AI should consider carefully how to balance competing interests and objectives for healthcare services delivered to similarly situated individuals and/or subgroups of individuals, especially if there is a risk of underrepresentation. Since AI must promote equal opportunities in terms of access to goods, services, and technology (European Commission 2019) in order to be equitable, it should not exacerbate existing health inequalities but facilitate and improve access to high-quality radiology services for all individuals and groups.

The central aspect of a fair approach to the use of AI in the treatment of images lies in the construction of a training dataset that is capable of representing patients in a way that does not hide common biases based on gender, ethnicity, income, or geographical origin. From the juridical point of view, the errors that could be derived from the training of AI algorithms become biased toward underrepresented groups or develop some discriminatory characteristics of the patient and, hence, exacerbate existing health disparities. It is thus central to the treatment of images with respect to references to personal data, since in the case of images, anonymization of the diagnostic imaging itself may not be sufficient to avoid identification by, for example, sex, when this may not be necessary for diagnosis, but, will in any case, only have a negative influence on the decision of the physician or radiologist.

On the other hand, personal attributes such as gender, age, ethnicity, and socioeconomic characteristics are not always maintained during data preparation, and the anonymization process in the preparation of images minimizes the possibility of patient identification when it is needed for certain medical predictions. As a matter of fact, the FUTURE-AI guidelines related to fairness should imply:

(i) The specific goals and end users of the AI solution. This is especially important in the case of sources of bias that are specific to the medical context.
(ii) The general conditions that should be satisfied in a horizontal manner and with respect to sex/gender, etc.

"The use of AI systems should never lead to people being deceived or unjustifiably impaired in their freedom of choice" (European Commission 2019) and should guarantee that "AI practitioners [respect] the principle of proportionality between means and ends" (European Commission 2019). These last two criteria in the Ethics Guidelines for the qualifications for fairness from the substantive point of view fall under the procedural dimension of fairness.

Fairness, when analyzed under the procedural dimension of the Ethics Guidelines, "entails the ability to contest and seek effective redress against decisions made by AI systems and by the humans operating them" (European Commission 2019). This is because "the entity accountable for the decision must be identifiable, and the decision-making processes should be explicable" (European Commission 2019).

These last substantive and procedural fairness considerations show the way to the other principles, especially explainability and accountability. Setting the elements for fairness is ontologically necessary in order to analyze the other principles, especially explicability.

5 The Relationship Between the GDPR and Images in the Context of Imaging Biobanks and Imaging Biomarkers

According to the definition of fairness, medical AI algorithms for imaging biobanks and imaging biomarkers should be built so that they address common as well as hidden biases, especially in training datasets, and are able to guarantee individual fairness and group fairness.

The problem of bias in AI is common to all medical applications; however, it is particularly relevant in medical imaging. To address the bias in AI for imaging biobanks and imaging biomarkers, we need to consider the consequences of the anonymization of diagnostic imaging as well as the use of images from clinical examinations, which qualify as personal data because they can be associated with the personal data of the patient (pseudoanonymized personal data). Both can create some bias in the light of the definition of fairness. It can be said that pseudoanonymized data have their own strong importance in the construction of the dataset to train the algorithm. At the same time, the anonymization of images, which is often sought in the construction of imaging repositories to avoid the application of the GDPR, does not always guarantee diagnostic efficiency because of the patient's interests and the research interest that also characterize biobanks of images. Thus, because of the nature of an imaging biobank and the imaging biomarkers created by integrating and collecting digital content from hospitals involved in a biobank, the relationship between the GDPR and the images must first be understood. Since our focus is on biobanks for medical imaging, the abovementioned relationship must logically be discussed before we consider the legal nature of clinical images as defined by the GDPR.

Sensitive data is a special category of personal data that is subject to additional protections. According to the GDPR, biometric data constitutes a "sensitive" category of personal data. The GDPR states that the processing of biometric data for identification purposes is prohibited. Processing is only justified if the explicit consent of the relevant person exists, specific legal obligations apply, or the processing is required for reasons of public interest. The GDPR (Art. 4) defines biometric data as "personal data resulting from specific technical processing relating to the physical, physiological or behavioural characteristics of a natural person, which allow or confirm the unique identification of that natural person, such as facial images".

To understand the position of clinical images – a part of a patient's medical record – we can treat them as photographs, and also as special photographs. According to recital 51 to the GDPR:

> the processing of photographs should not systematically be considered to be processing of special categories of personal data as they are covered by the definition of biometric data only when processed through a specific technical means allowing the unique identification or authentication of a natural person.

Therefore, the processing of images should not be systematically considered as the processing of a special category of personal data. Clinical images only fall within the definition of biometric data if they are processed by a specific technical means which allows for the unambiguous identification or authentication of a natural person.

Biobanks for medical imaging pseudonymize or anonymize the personal data of the patient.

In the context of biobanking for research, it is necessary to consider the possible reidentification of a pseudonymized or anonymized biospecimen. In the case of research projects in particular,[5] image anonymization is used. The difficulty of carrying out and guaranteeing anonymization is not analyzed here, but is just mentioned as an introduction to some of the problematic issues affecting the fairness of the use of AI in biobanks of images. Recital 26 of the GDPR defines the conditions for anonymization:

> The principles of data protection should apply to any information concerning an identified or identifiable natural person. Personal data which have undergone pseudonymization, which could be attributed to a natural person by the use of additional information should be considered to be information on an identifiable natural person. To determine whether a natural person is identifiable, account should be taken of all the means reasonably likely to be used, such as singling out, either by the controller or by another person to identify the natural person directly or indirectly. To ascertain whether means are reasonably likely to be used to identify the natural person, account should be taken of all objective factors, such as the costs of and the amount of time required for identification, taking into consideration the available technology at the time of the processing and technological developments. The principles of data protection should therefore not apply to anonymous information, namely information which does not relate to an identified or identifiable natural person or to personal data rendered anonymous in such a manner that the data subject is not or no longer identifiable. This Regulation does not therefore concern the processing of such anonymous information, including for statistical or research purposes.

Anonymization is an irreversible process applied to personal data to prevent the reidentification of the data subject. Consequently, the basic prerequisite is that the personal data was collected and processed, in compliance with the applicable legislation on the retention of data, in an identifiable manner. The notion of reasonable efforts on the part of the data controller or a third party requires the following to be taken into account:

(i) The technical tools at hand, bearing in mind possible changes in the development of Information and Communication Technologies (ICT).
(ii) The widest extent of data aggregation must be guaranteed. As a consequence, the reduction of immediately identifiable features is not in itself enough to avoid the identification of the data subject. It will probably be necessary to take additional measures to avoid identification, always depending on the specific circumstances and scope of the processing for which the anonymous data are to be used.
(iii) Third parties need to consider the risks of reidentification. This includes the situation in which a research project receives anonymized data and then has to perform an extra procedural check to address the risks of identification, connection, and implication.

In any case, reidentification via comparison with reference material may remain theoretical since the access to and the amount of the reference material (whole biobank archives), the personnel required, and the equipment and finances to perform such analyses would be immense. Even in the case of a medical imaging biobank, images of clinical examinations commonly qualify as personal data because they can be associated with the patient's personal data, albeit in a very indirect manner. This applies to images in biobanks in particular, although it is a common trap into which doctors and health specialists could fall in the light of recital 51. As a matter of fact, recital 26 GDPR does not make it clear that pseudonymized data should not be considered personal data in the hands of an entity that does not possess the code to reidentify the data, as long as appropriate measures are put in place to prevent that entity from gaining access to the key or otherwise linking the dataset to other data sources that may permit reidentification.

An additional problem in the relationship between GDPR and images is the research exemption. The basic prerequisite is that the personal data was collected and processed in compliance with the applicable legislation on the retention of data in an identifiable manner. Despite there being no mention in the GDPR that the determination of pseudonymization is a matter under the possible derogations of Member State law, the reidentification of anonymized personal data is one of the problems that must be solved in the light of national law, according to the possible different approaches at that level and taking into consideration the different positions of the data protection authorities at the national level. At the moment, the abovementioned problem can only be solved on a case-by-case assessment for each application. The main issue is related to the fact that the various processes and technologies for data anonymization seem to be able to generate "presumed anonymous data" and not "anonymous data" that falls outside the application of the GDPR. Thus, the approach that has been used to treat anonymized data is linked to a risk evaluation that the:

> data may, after all, be personal now or in the significant time range that needs to be taken into consideration. If this risk is considered to be marginal, it seems like a reasonable approach to treat the data as *anonymous*. Where this risk is above marginal, however, controllers may be well-advised to treat the data as *personal*.
>
> (Bruegger 2021, 116)

Taking into consideration this legal framework for personal data, we will next examine the bias arising from the treatment of medical images, which has a direct impact on fairness.

6 Metadata Labelling for Fairness and the Appropriate Representation of Patient Groups in the Training of Datasets With Images

During the collection and preparation of imaging databases for the development and testing of new AI algorithms, standardized metadata and key variables (e.g., sex/gender, ethnicity, age, and gender identity) are used for the purpose of data collection. Personal attributes such as sex, age, ethnicity, and socioeconomic characteristics, which are essential in order to comply with the principle that there should be no direct or indirect discrimination as required by the standard, are, however, not always retained during the data preparation and image anonymization processes. Although this operation is basic to the image anonymization process and is capable of minimizing the possibility of patient identification, it can open the door to bias in the AI system with respect to possible differences in health both within and between human groups and subpopulations defined by sex, age, ethnicity, and socioeconomic status. One particular aspect should be taken into consideration because of the anonymization of images: the labelling of metadata and the direct connection between the attributes mentioned earlier and the concept of fairness. Additionally, anonymization is not only inefficient for the purpose of avoiding the biases described earlier but also does not fully guarantee that an anonymous image does not generate discrimination. In the case of a framework suitable for testing for discrimination in AI systems, a recent early stage research study reported evidence that AI algorithms can identify race from a person's medical scans (e.g., chest and hand X-rays and mammograms), even when human medical experts would not be able to do this from the images without other references to personal data. In this particular case, the algorithmic and human discrimination problems, although different in nature and character, would have an impact on the principles of explicitness and fairness. This situation could generate a decision that is said to be based on AI but, in fact, is based on the different and unequal treatment of patients according to their race (Seyyed-Kalantari et al. 2021), even in cases in which the latter is not a relevant medical treatment criterion. In addition, the characterization of people during data collection and structuring is relevant.

As a matter of fact, qualitative biases affecting fairness may consist of the cognitive biases of the clinicians generating, interpreting, or annotating the imaging data (Geis et al. 2019), and thus understanding the human component in an imaging biobank is useful at several points, especially the accountability of decisions. Pseudonymization should be applied instead, and this process should include all the elements that define fairness. In this case, metadata that allows the measurement of the similarity of medical situations to verify the equal treatment of all similar cases (e.g., clinical information) is obtained from the process of labelling. It is important in imaging datasets to label the metadata, such as sex, gender, ethnicity, skin color, socioeconomic characteristics, and geography, to develop fair AI solutions. This activity must also be in compliance with data processing regulations. A labelling operation that is not conducted in compliance with the GDPR could undermine fairness. As a matter of fact, key areas of data labelling include informed consent, privacy and data protection, ownership, objectivity, transparency, the gap between those who have and those who do not have the resources to handle large datasets, and the provision of meaningful rights of access to the data (Geis et al. 2019).

7 Training AI Imaging Solutions Under European Non-Discrimination Law

The main cause of algorithmic bias in AI is the fact that training datasets often lack the quantitative and qualitative diversity and balance needed to achieve AI solutions that maintain the same performance across human groups and subpopulations and when they are applied to similarly situated individuals. As stated in the FUTURE-AI report reflections of this kind are also common in non-EU countries, and this brings together the experiences of leading radiology societies such as the American College of Radiology (ACR), the European Society of Radiology, the Radiological Society of North America (RSNA), the Society for Imaging Informatics in Medicine, the European Society of Medical Imaging Informatics, the Canadian Association of Radiologists, and the American Association of Physicists in Medicine (Geis et al. 2019). The fair representation of patient groups is to be achieved from both the medical point of view and in the light of sex, gender, ethnicity, geography, and socioeconomics, when and if those elements have an impact on existing health disparities. The "existing imaging databases are often imbalanced according to sex, ethnicity, geography and socioeconomics" (Lekadir et al. 2021). For this reason, there is a risk that trained AI algorithms will become biased with respect to underrepresented groups and, hence, exacerbate existing health disparities; because of this, AI tools may generate undetected errors with harmful consequences to the patient when they are applied in imaging conditions that differ or unexpectedly deviate, even slightly, from those used for training (Lekadir et al. 2021).

For the avoidance of doubt, this chapter talks about "sex" and not just about "gender" (see also chapter "A Possible Risk Governance Approach for AI in Health Research and Biobanking" by Kaya Akyüz, Mónica Cano Abadía, and Melanie Goisauf). This is because, from the legal point of view, "gender" refers to sexual orientation and identity: "people belonging to each of these categories are part of social . . . groups, created over time by the stratification of the social and legal rules that produce discrimination" (Billotta 2019, 369). Sexual orientation and identity do not have any relevance in the imaging used for medical treatment and diagnostics. Otherwise, to be "a woman or a man" could have some impact on the imaging used for medical treatment and diagnostics, because some medical conditions affect women and men differently (Marzo-Ortega et al. 2022) and knowing if a patient is "a woman or a man" could be relevant. The imaging used for medical treatment and diagnostics, of course, has no relevance to sexual orientation and identity. If an AI algorithm is trained with imaging data that is unbalanced with respect to sex (Larrazabal et al. 2020), socioeconomic characteristics, or ethnicity (Puyol-Antón et al. 2021), and there are differences in health within and between these groups, it is likely that the resulting model will produce biased predictions.

The FUTURE-AI guidelines offer – among other things – the solution of appealing for a multicenter data collection that can increase the diversity of datasets and avoid problems linked to the AI algorithm training with respect to sex, socioeconomic characteristics, or ethnicity, especially given the health differences within and across these groups. It is also possible to suggest that the AI imaging solution for distant mode training uses the so-called synthetic data. Since personal and nonpersonal data in a dataset are inextricably linked, synthetic data could support training.

Although multicenter data collection is a good solution if it is guided by a prior selection of represented groups, it is not certain that a perfect balance between different groups can be achieved in AI training, and, in the testing of datasets, the way to ensure compliance with the principle of fairness and non-discrimination is established by EU law and the EU courts (European Union Agency for Fundamental Rights 2010). Simple multicenter data collection might not be enough to avoid bias in the assessment of potential discrimination because particular statistical elements might not be satisfactory. Defining one or more disadvantaged groups, requires the making of case-specific normative choices that reflect the local political, social, and legal dimensions of the case (see chapter "The Approach of EU Law to Tackling Statistical Bias in Medicine" by Katharina Ó Cathaoir).

EU law gives guidance as to the specific fairness settings and general fairness requirements that should be met, as well as the requirements for specific clinical objectives, using the risk approach to select the fairness requirements that should be met.

> European non-discrimination law addresses two general types of discrimination: direct and indirect. Direct discrimination refers to adverse treatment based on a protected attribute such as sexual orientation or gender. Indirect discrimination on the other hand describes a situation where "apparently neutral provision, criterion or practice" disproportionately disadvantages a protected group in comparison with other people.
>
> (Wachter et al. 2021)

European non-discrimination law exists in both primary and secondary laws. The substantive dimension of fairness is mainly related to equal treatment under Article 19 of the Treaty on the Functioning of the European Union (TFEU), Article 21 of the EU Charter of Fundamental Rights (EUCFR), and Article 14 of the European Convention on Human Rights (ECHR), as well as EU non-discrimination law.

Turning to secondary law, four non-discrimination directives apply to both the private and the public sectors: the Racial Equality Directive (2000/43/EC), the Gender Equality Directive (recast) (2006/54/EC), the Gender Access Directive (2004/113/EC), and the Employment Directive (2000/78/EC). These non-discrimination directives only create a minimum standard and provide a general framework that needs to be transposed into the legal systems of the Member States, and they thereby introduce a fragmented standard for non-discrimination regulation across the Member States.

In the EU law system, there is a lack of clear EU legislative guidance to accommodate contextual equality in determining the correct representation of patient groups in the training of AI systems. The legal protection offered by the non-discrimination law is called into question when it is AI – not humans – that discriminates:

> Humans discriminate due to negative attitudes (e.g. stereotypes, prejudice) and unintentional biases (e.g. organisational practices or internalised stereotypes) which can act as a signal to victims that discrimination has occurred. Equivalent signalling mechanisms and agency do not exist in algorithmic systems. . . . The increasing use of algorithms disrupts traditional legal remedies and procedures for detection, investigation, prevention, and correction of discrimination which have predominantly relied upon intuition.
>
> (Wachter et al. 2021)

In this light, the discrimination of AI systems, as compared to traditional forms of discrimination, "is more abstract and unintuitive, subtle, intangible, and difficult to detect" (Wachter et al. 2021). We must recall the way in which discrimination could occur in imaging biobanks: artificial intelligence tools can generate undetected errors with harmful consequences for the patient, when applied to imaging conditions that may unexpectedly or even slightly deviate from those used for training. The EU legal system is the road map to use to approach the abstract and unintuitive discrimination in AI systems, through judicial interpretation and the evidential requirements used by the European Court of Justice (ECJ), even though these are difficult to apply and reconstruct, as suggested by some scholars (Wachter et al. 2021).

If not correctly represented, elements such as sex and race, understood as geographic origin, can have consequences for medical imaging: think of an AI system trained with the physical characteristics of Asian groups alone being applied in other geographical areas in which the AI tools have to analyze images in which the height of the patient has some medical consequence (Banerjee et al. 2021).

In this case, to understand whether the images and data used to train an algorithm are fair or not, we can use the conceptual map that the ECJ uses in deciding a case concerning alleged direct or indirect discrimination under EU non-discrimination law, as suggested by the abovementioned study (Wachter et al. 2021). Thus, the point is trying to apply the ECJ approach to prima facie discrimination. To base a claim on prima facie discrimination, claimants must prove that (1) a specific injury has happened or is upcoming (Case C-7/12); (2) the injury is revealed or is upcoming in a significant way within a protected group of persons (Joined Cases C-399/92, C-409/92, C-425/92, C-34/93, C-50/93, and C-78/93); and (3) the detriment for this group is disproportionate to the detriment for others in a similar situation (Case C-104/09).[6] For example, under the Racial Equality and Employment Directives, a comparison must always be made between the effects of the contested measure on two groups (Case C-167/97; Wachter et al. 2021, 15). At the same time, the judge must decide whether the disadvantaged group has been or is likely to be significantly disadvantaged compared to another group in a similar situation (Case C-167/97) and, thus, whether there is prima facie discrimination (Case C-127/92).

In the case of AI algorithm training, we have to move from the notion of plaintiff and judge to technicians for imaging biobanks in support of legal practitioners. The approach to the evaluation of an AI system in the EU legal system could be realized by the evidential requirements used by the ECJ to create a static baseline fairness metric in keeping with the "gold standard" set by the ECJ for assessing potential discrimination (Case C-167/97), which should be set by technicians for imaging biobanks in support of legal practitioners in a cross-disciplinary collaboration.

> Across these cases the ECJ and national courts have established a 'flexible and pragmatic' test for comparing effects on the disadvantaged and comparator group. To meet a European legal standard any attempt to automate judgements of discrimination would need to accommodate this flexibility and ideally facilitate the judiciary's 'gold standard' for full comparisons between disadvantaged and advantaged groups.
>
> (Wachter et al. 2021, 41)

The reference to the "gold standard" of the Court of Justice for assessing discrimination reduces the risk of subjectivity in the specific challenges arising from equity in the context of AI imaging solutions. In the conclusions of the FUTURE-AI guidelines related to fairness, there is mention of a subjective notion by nature. From a legal point of view, it is better not to talk about a "subjective notion" for "fairness", because something that could be defined as "subjective" is based on personal opinion and feelings rather than facts. It must be made clear that within the EU legal framework, any kind of AI system must be based on the hierarchy of EU legal sources, and nothing else. In the hierarchy of EU legal sources that applies to AI, this essentially means "fundamental rights". According to the Ethics Guidelines or the various documents on this topic, the fundamental rights are the basis on which AI can realize the meaning of respect for fundamental rights within the framework of democracy and the rule of law, and they provide the most extensive values that can be made operational in AI systems. The prima facie "gold standard" of the Court of Justice and the multidisciplinary approach must be used to reduce subjectivity, because they are able to identify as many sources of application-specific bias and inequity as possible through the interaction between medical doctors, the technical community, and social scientists, ethicists, and, especially, jurists. What needs to be avoided is a situation in which system developers and controllers alone set regulatory thresholds for discrimination at the local and subjective levels without external regulatory or judicial input (Wachter et al. 2021).

8 AI Tools, the Interpretation Skills of Experienced and Less Experienced Radiologists, and Fairness

Overconfidence in the outcomes of the use of AI support may affect fairness, particularly in relation to diagnostic images and, especially, the interpretation of mammograph images.[7] AI tools affect the decision-making and interpretation skills of experienced and less experienced radiologists. It is well known in the medical literature that the integration of AI in real-life clinical practice can raise issues of fairness, including when it is used by experienced and less experienced physicians, and this can have an effect on their decision-making abilities. For example, in relation to the interpretation of mammograph images, one study by Rauscher et al. (2013) has reported how AI-automated support positively influenced the decision-making of radiologists with less advanced interpretative skills. The same study described how AI systems had a negative effect on the decision-making of radiologists with advanced image interpretation skills.

This is evidently related to overconfidence in the outcome of automated support and a devaluation of physicians'/medical doctors' professional experience accumulated over years of medical practice. It should also be considered that this may be dictated by misconceptions about the limitations and the strengths of AI.

To avoid the bias derived from overconfidence in the outcome of automated support and the devaluation of professional experience, "people involved with any stage in an AI product's life cycle must understand it deeply" (Geis et al. 2019). As a matter of fact,

> only the correct and complete identification of the people involved in the process of creating and using biobanks and their specific roles allow to correctly defining the legal aspects that must be respected in this field. The proper application of regulation is the fundamental requirement of the whole processes.
>
> (Burzagli 2023)

"Adequate information and training material should be provided to raise awareness and inform end-users on the fairness, biases and limitations of the AI algorithm" (Lekadir et al. 2021). The professional people involved in the life cycle of an AI product have to be informed about and trained in the meaning of fairness to avoid the biases derived from overconfidence in the outcomes of automated support. The training must enable them to recognize that, while AI systems bring substantial benefits to individuals and society, they also present risks of negative impacts, including impacts that are difficult to predict, identify, or measure. The EU uses an approach based on fundamental rights, which means fundamental rights are respected within the framework of democracy and the rule of law, and the most extensive values that can be made operational in an AI system are provided. These values are related not only to the fairness of AI systems (because they affect the rule of law) but also to distributive justice (if we think of access to healthcare) and, in the case of radiology, they affect the human mind itself. Measures should be taken to mitigate these risks when appropriate and in proportion to their magnitude.

9 Conclusion

In this chapter, we reflect on the meaning of fairness in the life of imaging biobanks and in the application of AI systems to develop imaging biomarkers. To reach this goal, we analyze the report by Lekadir et al. (2021) titled "FUTURE-AI: Guiding

Principles and Consensus Recommendations for Trustworthy Artificial Intelligence in Medical Imaging"[8] because these guidelines deal with cases and experience as well as offering solutions for realizing an equitable AI system in medical imaging. We use this report because it reproduces the principles and related recommendations derived from the five largest European projects on AI in health imaging (Section 4). The experiences from these projects could be defined as best practices for approaching the common problems arising from the use of AI in imaging biobanks and in the development of imaging biomarkers. Focusing on fairness and its legally relevant aspects is very useful, not least because the elements derived from an analysis of fairness have a direct impact on the other principles of universality, traceability, usability, robustness, and, in particular, explicability.

According to the FUTURE-AI guidelines, explainability for medical imaging means that

> medical AI algorithms should be able to provide meaningful and actionable explanations to the clinicians for their predictions. Explainability provides insight into the algorithmic mechanisms behind AI decision making processes thereby allowing for clinical validation and scrutinization of these decisions. While local explainability highlights the reasons behind a particular prediction by the AI model for an individual image, global explanations identify the common characteristics that the AI model considers important for a particular image analysis task. Attribution maps (or heat-maps) are commonly used visual methods for explainability in medical AI, which highlight the relevant regions on the input image that the AI model considers important.
>
> (Lekadir et al. 2021)

Explainability "speaks" to different groups of "people" and has different roles in the process of verifying the application of the requirements for AI so that people are informed and trained to ensure fairness (see Sections 6 and 7). The understanding at a specific stage of the process of who is the end user in a biobank opens up the question of how to explain decision-making processes. If the end user is represented by a patient/data subject who uses a treatment based on the results produced by AI tools (Burzagli 2023), then – in the case of an imaging biobank – the explanation of the use of AI to treat the health personal data of the research participant/patient/data subject must be given to: "Persons who develop the requirements, people who use the systems, and end-users, who, for example, in the healthcare, can be represented by patients who use treatment based on the results produced by Artificial Intelligence tools" (Burzagli 2023).

In the specific context of biobanks, a further element is added to the classification expressed earlier: the difference between supplier and user is inserted. The supplier is the person who allows the collection of his or her own information, whether physical or not. Each sample, even a physical sample, corresponds to a profile of the person to whom that sample belongs. The user or end user is the person who takes advantage of the processing of data stored in the biobank (Burzagli 2023).

The research participant/patient/data subject is often taken to be a single person, but she/he also has a "social dimension, which is represented by the community of people who generally use the output of processing of data contained in the biobanks" (Burzagli 2023). Biobanking activities are vital research tools in the medical sciences, and this has led to a widespread debate in the literature about how best to handle the consent procedures governing the enrolment of participants in research as end users, and the subsequent use of participant imaging and data in other studies. Research and other activities carried out within a biobank may not be consistent with the right of the individual to give his/her consent to the processing of personal data. In this sense, it is fundamental to define broad consent and/or dynamic consent in the light of future research.

The other point that arises from the correct analysis of fairness is the principle of traceability. This principle states that medical AI algorithms should be developed in conjunction with mechanisms to document and monitor the entire development life cycle as well as the functioning of AI tools in their environment. These aspects increase transparency by providing detailed and comprehensive information on, for example, the datasets used to train and evaluate the algorithms, the associated protocols and parameters, the variables included in the AI models, and any potential biases and limitations, not only for biobank operators (to assess the performance and limitations over time and, thus, to identify potential sources of error or drift from training settings or previous states) but especially for research participants/patients/data subjects. The latter thus see the principles derived from the Helsinki Convention reinforced in the requirement to give their consent to be involved in the biobank research activity.

The correct interpretation of the legal and ethical aspects of fairness also has an impact on the principle of "universality". This standard, which includes technical, clinical, ethical, and regulatory standards, has to achieve three key objectives:

(i) Enabling the development of artificial intelligence technologies with greater interoperability and applicability across clinical centers, radiology units, and geographical locations.
(ii) Fostering a culture of quality, safety, and trust in medical AI based on well-established and widely accepted frameworks.
(iii) Facilitating co-creation and cooperation in medical AI among AI developers, manufacturers, radiologists, physicians, data managers, and healthcare entities based on a unified language and common approaches.

At the same time, the principle of robustness is derived from the prerequisites of fairness because it refers to the ability of a medical AI model to maintain its performance and accuracy when applied under highly variable conditions in the real world, outside the controlled environment of the laboratory where the algorithm was built. The principle of robustness is rooted in the need for fairness, as it emphasizes a medical AI model's ability to maintain performance and accuracy in diverse real-world conditions, beyond the controlled laboratory environment where the algorithm was developed. It is crucial for AI algorithms to be designed with human factors in mind for each clinical task. This can be achieved through human-centered approaches that involve user engagement throughout the AI development process, including iterative usability testing, referred to as the Usability Principle. By focusing on these human-centered methodologies and maintaining user involvement, developers can ensure that AI algorithms effectively meet the needs of clinical practice.

The robustness of a model is its ability to generalize and predict well even in the presence of changing conditions that cause domain changes and with datasets that may or may not have been predicted prior to implementation. This element has a direct link to the formation of the dataset for training the AI tools. From a legal point of view, analyzing fairness is the main point in ensuring trustworthy AI, especially in the field of medical images. The better approach to avoid the bias from the use of AI in medical imaging that is derived from retaining, or not retaining, personal attributes such as sex, age, ethnicity, and socioeconomic characteristics in the images (as explained in Section 6) is to consider the external regulatory or judicial inputs. This will prevent AI system developers and controllers themselves from establishing references to normative thresholds for discrimination at a subjective level.

The EU legal system is the road map for dealing with abstract and nonintuitive discrimination in AI systems, but in order to prevent a dutiful reference to fundamental rights and, with it, the rules on discrimination becoming generic and difficult to implement, the judicial interpretation of European non-discrimination law has to be taken into consideration, as well as the evidence requirements used by the Court of Justice of the European Union in prima facie judgments to assess the points to focus on for discrimination. Reference to the Court of Justice's prima facie "gold standard" for assessing discrimination reduces the risk of subjectivity in specific equity challenges in the context of AI imaging solutions by creating a static baseline fairness metric. Interdisciplinary collaboration between jurists, ethicists, medical radiologists, and AI system developers engaged in the specific case of biobank imaging and biomarker development is the way to identify as many sources of application-specific bias and inequality as possible (see also Lekadir et al. 2025), and to attempt to develop an approach in line with the "gold standard" established by the Court of Justice for assessing potential discrimination in diagnostic imaging.

Acknowledgments

This project has received funding from the European Union's Horizon 2020 research and innovation program under grant agreement No 952159 (ProCAncer-I).

Notes

1 The issue of the use of algorithms and AI in general is the subject of documents, such as the so-called AI Act; as a matter of fact, on 21 May 2024, the Council of the European Union approved the EU Artificial Intelligence Act (AI Act). This is the final stage in the legislative process and comes after the EU Parliament voted to adopt the legislation

on 13 March 2024. In any case, the AI Act is the most important but not the only EU document related to AI. See, the Resolution of the European Parliament of 16 February 2017 with recommendations for the Commission on "Civil Law Rules on Robotics"; the Communication of the European Commission of 25 April 2018 on "Artificial Intelligence for Europe" (COM(2018) 237 final); the Communication of the European Commission of 8 April 2019 "Building Trust in Human-centric Artificial Intelligence" (COM(2019) 168 final), and, more recently, the White Paper "On Artificial Intelligence – A European Approach to Excellence and Trust" of 19 February 2020. Within the framework of the Council of Europe, the Committee of Experts on Internet Intermediaries published the study "Algorithms and Human Rights" in March 2018, and shortly afterward, on 3 December 2018, the European Commission for the Efficiency of Justice Systems (CEPEJ) approved the "Ethical Charter on the Use of Artificial Intelligence in Justice Systems and their Environment". All this has resulted in the European Commission's Communication called "A European Strategy for Data" (COM(2020) 66 final), published on 19 February 2020. It is part of a wider package of strategic documents, including a Communication on "Shaping Europe's Digital Future" and, of course, the aforementioned White Paper "On Artificial Intelligence – A European Approach to Excellence and Trust". New interest in the legal and ethical features of artificial intelligence has been triggered by the publication of important documents at the EU level by the European Commission's High-Level Expert Group on Artificial Intelligence (AI HLEG–the European Commission appointed a group of experts to provide advice on its artificial intelligence strategy: https://digital-strategy.ec.europa.eu/en/policies/expert-group-ai). The panel's recommendations were used as inputs for policy initiatives adopted by the EU Commission and the EU Member States, such as the Communication on "Building Trust in Human Centric Artificial Intelligence", the White Paper on "Artificial Intelligence: A European Approach to Excellence and Trust", and the updated "Coordinated Plan on AI".

2 See, in Europe, the ESFRI (European Strategy Forum for Research Infrastructures) road map for research infrastructures, which lists the research infrastructure for biobanks and biomolecular resources (BBMRI). https://roadmap2021.esfri.eu/projects-and-landmarks/browse-the-catalogue/bbmri-eric/. (08.04.2025).

3 See the BBMRI.it definition at www.bbmri.it/en/nodo-nazionale/biobanche/. (08.04.2025).

4 For example, the Pro-Cancer AI project (European Union's Horizon 2020 research and innovation program under grant agreement No 952159) used anonymization for the treatment of health personal data and images, focused on missed imaging appointments in the United States, and found that racial minorities and people with low socioeconomic status missed more healthcare appointments than other groups, which further contributes to their prevailing underrepresentation in the imaging datasets of their healthcare providers.

5 For example, the Pro-Cancer AI project (European Union's Horizon 2020 research and innovation program under grant agreement No 952159) used anonymization for the treatment of health personal data and images.

6 In this case, the ECJ accepted as unlawful discrimination the unfavorable treatment of a subcategory of men.

7 In relation to breast cancer, women with high breast density are generally underrepresented in the existing clinical registries, see Lekadir et al. (2021).

8 The work of the consortium continued. Over two years, the FUTURE-AI guideline was developed through consensus on six key principles: fairness, universality, traceability, usability, robustness, and explainability. To ensure trustworthy AI in healthcare, 30 best practices were established, covering technical, clinical, socio-ethical, and legal aspects across the AI life cycle, from design to deployment and monitoring. See Lekadir et al. 2025.

Reference List

Banerjee, Imon, Ananth Reddy Bhimireddy, John L. Burns, et al. 2021. "Reading Race: AI Recognises Patient's Racial Identity." *Medical Images*: 10356.

Barucci, Andrea. 2022. "The Era of Artificial Intelligence in Precision Medicine and Beyond." *Working Paper Presented at Seminario Internazionale Sull'intelligenza Artificiale*, University of Perugia May 19 and 20.

BBMRI.it. n.d. "Biobanks." www.bbmri.it/en/nodo-nazionale/biobanche/ (09.04.2025).

Billotta, Francesco. 2019. "LGBTIQ People." In *Dictionary of Statuses within EU Law*, edited by A. Bartolini, R. Cippitani and V. Colcelli. Cham: Springer Nature Switzerland.

Borgheresi, Rita, Andrea Barucci, Sara Colantonio, et al. 2022. "NAVIGATOR: An Italian Regional Imaging Biobank to Promote Precision Medicine for Oncologic Patients." *European Radiology Experimental* 6 (1): 53. https://doi.org/10.1186/s41747-022-00306-9.

Bruegger, Bud P. 2021. "Towards a Better Understanding of Identification, Pseudonymisation, and Anonymisation." *Unabhängiges Landeszentrum für Datenschutz Schleswig-Holstein (ULD)*. www.datenschutzzentrum.de/uploads/projekte/IdentPseudoAnon-320-v1-0-web.pdf (09.04.2025).

Burzagli, Laura. 2023. "In the Context of Biobanks Who Are the Users and Which Factors Relating to the Human Component to Consider?" In *GDPR Requirements for Biobanking Activities Across Europe Handbook*, edited by Rainer Arnold, Roberto Cippitani, Valentina Colcelli and Christoph Brochhausen. Cham: Springer Nature Switzerland.

Diaz, Oliver, Kaisar Kushibar, Richard Osuala, et al. 2021. "Data Preparation for Artificial Intelligence in Medical Imaging: A Comprehensive Guide to Open-Access Platforms and Tools." *Physica Medica* 83: 25–37. https://doi.org/10.1016/j.ejmp.2021.02.007.

European Commission. 2018. Communication of the European Commission of 25 April 2018 on Artificial Intelligence for Europe COM(2018) 237 final.

European Commission. 2019. Communication of the European Commission of 8 April 2019 on Building Trust in Human-centric Artificial Intelligence COM(2019) 168 final.

European Commission. Directorate-General for Communications Networks, Content and Technology. 2019. "Ethics Guidelines for Trustworthy AI." Publications Office.

European Society of Radiology. 2015. "ESR Position Paper on Imaging Biobanks." *Insights into Imaging* 6 (4): 403–10. https://doi.org/10.1007/s13244-015-0409-x.

European Union Agency for Fundamental Rights. 2010. *Handbook on European Non-Discrimination Law*. https://fra.europa.eu/sites/default/files/fra_uploads/1510-fra-case-law-handbook_en.pdf (04.09.2023).

Gaskell, George, Nick Allum, Martin Bauer, et al. 2000. "Biotechnology and the European Public." *Nature Biotechnology* 18 (9): 935–8. https://doi.org/10.1038/79403.

Geis, J. R., A. P. Brady, C. C. Wu, et al. 2019. "Ethics of Artificial Intelligence in Radiology: Summary of the Joint European and North American Multisociety Statement." *Radiology* 293 (2): 436–40. https://doi.org/10.1148/radiol.2019191586.

Glover, McKinley, Dania Daye, Omid Khalilzadeh, et al. 2017. "Socioeconomic and Demographic Predictors of Missed Opportunities to Provide Advanced Imaging Services." *Journal of the American College of Radiology* 14 (11): 1403–11. https://doi.org/10.1016/j.jacr.2017.05.015.

Hacker, Philipp. 2018. "Teaching Fairness to Artificial Intelligence: Existing and Novel Strategies Against Algorithmic Discrimination Under EU Law." *Common Market Law Review* 55: 1143–86.

Larrazabal, A. J., N. Nieto, V. Peterson, D. H. Milone, and E. Ferrante. 2020. "Gender Imbalance in Medical Imaging Datasets Produces Biased Classifiers for Computer-Aided Diagnosis." *Proceedings of the National Academy of Sciences of the United States of America* 117 (23): 12592–4. https://doi.org/10.1073/pnas.1919012117.

Lekadir, K., A. F. Frangi, A. R. Porras, B. Glocker, C. Cintas, C. P. Langlotz, E. Weicken, F. W. Asselbergs, F. Prior, G. S. Collins, G. Kaissis, G. Tsakou, I. Buvat, J. Kalpathy-Cramer, J. Mongan, J. A. Schnabel, K. Kushibar, K. Riklund, K. Marias, L. M. Amugongo, L. A. Fromont, L. Maier-Hein, L. Cerda-Alberich, L. Marti-Bonmati, M. J. Cardoso, M. Bobowicz, M. Shabani, M. Tsiknakis, M. A. Zuluaga, M. C. Fritzsche, M. Camacho, M. G. Linguraru, M. Wenzel, M. De Bruijne, M. G. Tolsgaard, M. Goisauf, M. Cano Abadia, N.

Papanikolaou, N. Lazrak, O. Pujol, R. Osuala, S. Napel, S. Colantonio, S. Joshi, S. Klein, S. Ausso, W. A. Rogers, Z. Salahuddin, M. P. A. Starmans, and Future-AI Consortium. 2025. "FUTURE-AI: International Consensus Guideline for Trustworthy and Deployable Artificial Intelligence in Healthcare." *BMJ* 388: e081554. https://doi.org/10.1136/bmj-2024-081554.

Lekadir, Karim, Richard Osuala, Catherine Gallin, et al. 2021. "FUTURE-AI: Guiding Principles and Consensus Recommendations for Trustworthy Artificial Intelligence in Medical Imaging." *CoRR*. https://doi.org/10.48550/arXiv.2109.09658.

Marzo-Ortega, H., V. Navarro-Compán, S. Akar, U. Kiltz, Z. Clark, and E. Nikiphorou. 2022. "The Impact of Gender and Sex on Diagnosis, Treatment Outcomes and Health-Related Quality of Life in Patients with Axial Spondyloarthritis." *Clinical Rheumatology* 41 (11): 3573–81. https://doi.org/10.1007/s10067-022-06228-6.

Mitrou, Lilian. 2018. "Data Protection, Artificial Intelligence and Cognitive Services: Is the General Data Protection Regulation (GDPR) 'Artificial Intelligence-Proof'?" *SSRN Electronic Journal*. SSRN: https://ssrn.com/abstract=3386914 or http://dx.doi.org/10.2139/ssrn.3386914 (31.12.2018).

OECD. 2007. "OECD Best Practice Guidelines for Biological Resource Centres." *OECD Publishing*. https://doi.org/10.1787/9789264128767-en.

Park, Yoonyoung, Gretchen Purcell Jackson, Morgan A. Foreman, Daniel Gruen, Jianying Hu, and Amar K. Das. 2020. "Evaluating Artificial Intelligence in Medicine: Phases of Clinical Research." *JAMIA Open* 3 (3): 326–31. https://doi.org/10.1093/jamiaopen/ooaa033.

Perry, Mark. 2012. "Accessing Accessions: Biobanks and Benefit-Sharing." In *Comparative Issues in the Governance of Research Biobanks*, edited by Giovanni Pascuzzi, Umberto Izzo and Matteo Macilotti. Berlin, Heidelberg: Springer.

Povyakalo, Andrey A., Eugenio Alberdi, Lorenzo Strigini, and Peter Ayton. 2013. "How to Discriminate Between Computer-Aided and Computer-Hindered Decisions: A Case Study in Mammography." *Medical Decision Making* 33 (1): 98–107. https://doi.org/10.1177/0272989x12465490.

Puyol-Antón, Esther, Bram Ruijsink, Stefan K. Piechnik, et al. 2021. "Fairness in Cardiac MR Image Analysis: An Investigation of Bias Due to Data Imbalance in Deep Learning Based Segmentation." In *Medical Image Computing and Computer Assisted Intervention – MICCAI 2021*, edited by de Bruijne et al. Cham: Springer. https://doi.org/10.1007/978-3-030-87199-4_39.

Rauscher, Garth H., Jenna A. Khan, Michael L. Berbaum, and Emily F. Conant. 2013. "Potentially Missed Detection with Screening Mammography: Does the Quality of Radiologist's Interpretation Vary by Patient Socioeconomic Advantage/Disadvantage?" *Annals of Epidemiology* 23 (4): 210–14.

Seyyed-Kalantari, L., G. Liu, M. McDermott, I. Y. Chen, and M. Ghassemi. 2021. "CheXclusion: Fairness Gaps in Deep Chest X-Ray Classifiers." *Pacific Symposium on Biocomputing* 26: 232–43.

Slokenberga, Santa. 2021. "Setting the Foundations: Individual Rights, Public Interest, Scientific Research and Biobanking. GDPR and Biobanking." In *GDPR and Biobanking*, edited by Santa Slokenberga, Olga Tzortzatou and Jane Reichel. Cham: Springer.

Staunton, Ciara, Santa Slokenberga, and Deborah Mascalzoni. 2019. "The GDPR and the Research Exemption: Considerations on the Necessary Safeguards for Research Biobanks." *European Journal of Human Genetics* 27 (8): 1159–67.

Wachter, Sandra, Brent Mittelstadt, and Chris Russell. 2021. "Why Fairness Cannot Be Automated: Bridging the Gap Between EU Non-discrimination Law and AI." *Computer Law & Security Review* 41 (2021): 105567. http://doi.org/10.2139/ssrn.3547922.

Zika, Eleni, Daniele Paci, Tobias Schulte In Den Bäumen, et al. 2010. *Biobanks in Europe: Prospects for Harmonisation and Networking*. Luxembourg: Publications Office of the European Union.

Legal Sources

CaseC-7/12 Nadežda Riežniece v Zemkopības ministrija and Lauku atbalsta dienests ECLI:EU:C:2013:410 [2013] ECR.

CaseC-104/09 Pedro Manuel Roca Álvarez v Sesa Start España ETT SA ECLI:EU:C:2010:561 [2010] ECR I-08661.

Case C-127/92 Dr. Pamela Mary Enderby v Frenchay Health Authority and Secretary of State for Health ECLI:EU:C:1993:859 [1993] ECR I-05535.

Case C-167/97 Regina v Secretary of State for Employment, Ex Parte Nicole Seymour-Smith and Laura Perez ECLI:EU:C:1999:60 [1999] ECR I-00623.

Joined Cases C-399/92, C-409/92, C-425/92, C-34/93, C-50/93 and C-78/93 Stadt Lengerich v Angelika Helmig and Waltraud Schmidt v Deutsche Angestellten-Krankenkasse and Elke Herzog v Arbeiter-Samariter-Bund Landverband Hamburg eV and Dagmar Lange v Bundesknappschaft Bochum and Angelika Kussfeld v Firma Detlef Bogdol GmbH and Ursula Ludewig v Kreis Segeberg ECLI:EU:C:1994:156 [1994] ECR I-05727.

Regulation (EU) 536/2014 on Clinical Trials on Medicinal Products for Human Use [2014] OJ L 158.

Regulation (EU) 2016/679 on the Protection of Personal Data of Natural Persons with Regard to the Processing of Personal Data and on the Free Movement of Such Data (General Data Protection Regulation) [2016] OJ L 119.

7 Artificial Intelligence and Communication With Research Participants in the Process of Biobanking Human Biological Material

*Jarosław Sak and Jakub Pawlikowski**

1 Introduction

Artificial intelligence (AI) includes the development of computerized models to perform tasks that usually require human intelligence and is primarily the domain of IT knowledge. AI algorithms are programmed to learn and identify patterns from "training data" that can then be applied to new datasets without being explicitly programmed for that purpose. One application of AI is machine learning (ML), which is the process of using mathematical data models to help a computer learn without direct instruction.

Currently, AI and ML have found increased applications in medical care and biomedical research, including in the biobanking of human biological material (HBM) for research purposes. AI is revolutionizing the field of medical imaging, and many solutions based on AI are applied in radiology, pathology (Kaneko et al. 2022), cardiology (Fotaki et al. 2022; Chen et al. 2020), and ophthalmology (Han et al. 2021; Urias et al. 2019). Over the past decade, AI has become a breakthrough technology in pharmaceutical research and development. While AI cannot replace human ingenuity, it is very useful in clinical trial design, conduct, and analysis. The COVID-19 pandemic accelerated the use of AI in clinical trials due to an increased reliance on digital technology in clinical trial conduct (Kolluri et al. 2022; Vijayan et al. 2022).

The applications of AI in medical devices, medical diagnosis, risk prediction, and treatment support are expanding rapidly and help healthcare professionals to diagnose, treat, and care for their patients in more effective and more personalized ways. From the perspective of personalized medicine, the biobanking of human biological material (HBM) supports the development of preclinical research and the translation of basic research into clinical applications according to the principles of prediction, prevention, and personalization of medicine (PPPM/3PM) (Kinkorová and Topolčan 2020).

AI offers new opportunities for research on HBM stored in biobanks. It may help answer several questions regarding the impact of individual factors on human health by integrating genomic and other omics data with electronic health record data or by developing phenotypes that are more precise than those supplied by billing codes for diseases (Narita et al. 2021). It may help to study the impact

DOI: 10.4324/9781032634487-11

of genetic variation on human health by integrating genomic data with electronic health record data, and the implications of genetic heterogeneity for drug efficiency or adverse drug reactions can also be traced or captured by machine learning (Narita et al. 2021). The utilization of massive complex datasets stored in biobanks that combine individual genes, functions, and environmental differences can improve diagnosis, develop new treatment approaches, reduce side effects, and predict prognosis. An AI system connected to an automated sample storage system may also relocate biosamples to maximize storage space utilization. International biobanking collaboration may help facilitate access to and analyze higher volumes of data and images (Kozlakidis 2020). In addition to using AI in research, analyses of biodata, and the management of samples storage processes, AI can be used to improve mutual communication with research participants at different stages of the biobanking process, beginning from the process of searching for potential participants, through the process of obtaining informed consent, and to the moment the communication of results and controlling future uses. The aim of the chapter is to analyze the role of AI in communication with participants at various stages of biobanking and to discuss the ethical and legal issues that arise in such circumstances. We also discuss potential future uses and applications of AI in biobanking[1] and the research on HBM.

Along with the many benefits connected with using AI in healthcare and research, this innovative technology also faces significant ethical, legal, and societal questions about how AI systems should be used, the levels of risks they involve, the responsibilities for errors, or how we can control these machines, which are currently being discussed vigorously (McCradden et al. 2022; Abdullah et al. 2021; Murphy et al. 2021; Gerke et al. 2020; Müller 2021). However, despite many ethical and legal issues related to the use of AI in medical research, it may also contribute to increasing respect for the autonomy and dignity of research participants. Overall, it may also increase public trust in biobanks.

2 Searching for Potential Participants

AI algorithms are normally used to process and manage big datasets, which are defined as "datasets whose size is beyond the ability of typical database software tools to capture, store, manage, and analyze" (Manyika et al. 2011). The use of this type of algorithm in the process of searching for potential biobanking participants in the digital world will be discussed. It precedes the first personal meeting with potential participants in a biobank or unit where samples and data are collected; therefore, we propose a new term for this stage: "pre-biobanking". This term should be understood as related to the stage of looking for potential participants. The activities at this stage may be extended by the possibilities of AI.

The use of AI algorithms in the process of searching for research participants according to specific features seems to be easily implemented because much research is based on recruiting participants through the Internet or social media (Klassen et al. 2018; Cahill et al. 2019). Social media allows users to communicate and exchange information about users. Adverts, also about research, can be posted and promoted to specific target audiences by demographics, such as region, age, or gender. There is growing evidence to suggest that Facebook and other social media platforms are useful recruitment tools, and their use should thus be considered when conducting future health research. Compared with traditional recruitment methods (print, radio, television, and email), the benefits of innovative recruitment tools include reduced costs, shorter recruitment periods, better representation, and improved participant selection in young and hard-to-reach demographics. However, it remains limited by Internet access and the overrepresentation of young white women (Whitaker et al. 2017). It should be noted that searching for potential participants via the Internet or social media raises many ethical and legal problems regarding privacy (e.g., personal data security and management, anonymity, and confidentiality), authenticity, informed consent, recruitment, voluntary participation, and minimizing harm (Hunter et al. 2018). However, it is possible to search for potential participants while respecting the privacy, autonomy, and voluntariness of social media users according to existing guidelines (Darko et al. 2022).

Searching for potential participants in biobanking requires considering who (which subpopulation) would be interested in biobanking and how we should communicate with them about biobanking and research on HMB. Then, an initial selection of individuals from the entire population based on the features that best characterize a specific population target is made. This may be achieved by appropriately prepared and targeted advertisements or messages posted on the Internet or social media. This could also be done through advertisements about the activity of a specific biobank targeted at defined groups, as is the case in popular advertisements among Internet users.

To successfully enroll new participants in biobanking using AI algorithms, it is crucial to recognize the social, demographic, and psychological factors associated with the willingness of individuals to participate in biomedical research. In a review of 61 previous studies in social research, it was found that donation is a complex process that may be determined by psychosocial factors, such as people's knowledge and positive opinions on biobanks, trust, beliefs about the expected benefits, access to the information about the research (who, where, and what was the purpose), donors' cultural and religious beliefs, and privacy protection (Domaradzki and Pawlikowski 2019). Several studies indicate that young and middle-aged persons are more favorable toward donation (Prictor et al. 2020; Rahm et al. 2013), as well as those with higher education (Gaskell et al. 2013) and higher economic status, those who live in urban areas, and those who have children (Ahram et al. 2014). In contrast, being a member of an ethnic minority is associated with lower willingness (Abu Farha et al. 2020). Psychology studies have revealed the role of social trust, social involvement, preferred value system, and pro-social attitude in the willingness to donate samples to biobanks (Broekstra et al. 2021). The results of a national study carried out in Poland in a group of 1,100 adult people indicate that the willingness to donate samples to a biobank was significantly positively associated with trust in doctors and scientists, the tendency toward risk, and the following preferred values: work, helping others, travels, personal development, tradition, and knowledge. The regression analysis indicates that the willingness to donate material to a biobank is best explained by trust in scientists and doctors and the preferred values of personal development and tradition (Pawlikowski et al. 2022). Using AI in the process of searching for potential research participants in biobanking and establishing communication with them may be a cost-effective and efficient strategy for reaching and recruiting research participants. This tool may be particularly useful in communicating with the younger generation. Additionally, psychosocial variables connected with social media use that characterize recruited participants may be included in biobank datasets (based on the informed consent and voluntary act of recruited participants). We have many examples of research based on samples and data stored in biobanks in which psychosocial features played a crucial role, such as the role of personality traits in smoking (Sallis et al. 2019), reporting adverse drug reactions (Realo et al. 2018), and the polygenic risks for schizophrenia (Escott-Price et al. 2019) and depression (Brailean et al. 2020; Polimanti et al. 2019; Ronaldson et al. 2022).

Another way to search for potential donors according to specific features is recruiting through existing medical databases. Currently, this has limitations due to the limited knowledge regarding which data may be useful and limited access to the desired parameters. However, adapted (mature) AI algorithms could work for selecting potential donors or data of, for example, retinal images recorded digitally during standard ophthalmic follow-up visits, digitized X-rays, or CT images of the chest from standard follow-up examinations, both from healthy people and those treated for chronic diseases. Retinal images could also be acquired and processed as part of general practitioner practice. Similarly, it would be possible to use many other types of data, especially laboratory and imaging data, obtained while conducting medical examinations for employees or students as part of occupational medicine and school medicine, respectively. In this context, a closer collaboration between biobanks, general practitioners, hospitals, and blood donation centers (the latter cooperate with many blood donors and have a lot of clinically valuable medical and epidemiological data) would be desirable. Having access to valuable clinical and epidemiological data, AI algorithms could select individuals whose clinical profiles are coherent with the characteristics of participants in already designed or ongoing experimental research or clinical trials. In this way, it would be possible to search for and engage participants purposely and more accurately. However, as not all hospitalized patients (or donors in blood centers) consent to having their data and samples shared with biobanks and researchers, the implementation of AI in the analysis of existing databases is a process that requires many legal and organizational changes and solutions enabling data sharing between hospitals (and other healthcare institutions) and biobanks. The key factor is the legal ability of healthcare institutions to share medical data for analysis by AI algorithms and patients/donors/participants' consent to be included in the search and to be contacted.

An interesting proposal is to establish a network and infrastructure of pre-biobanks and pre-research medical databases, which, by processing the collected data, would create sets of potential participants to whom invitations to donate biological samples or additional data should be addressed. For this purpose, legal solutions are needed regarding extending the consent of the patients to the use of their material and data in the pre-biobanking process, taking into account the existing regulations in national and international laws. The establishment of such pre-biobanking institutions would be financially effective. The existing nationwide prospective genomic cohorts that store a variety of biological samples, such as blood, urine, and tissues, do not always achieve full cost-effectiveness, and the redundancy of the derived data can sometimes cause significant difficulties in their processing. Therefore, it is worth considering the possibility of preselection of data obtained by healthcare institutions and data collected by biobanks. Pre-biobanking elements were included in the PRIMAGE project, which aims to establish an imaging biobank in pediatric oncology (PRIMAGE n.d.). The project includes tasks related to the interoperability between biobanks that collect and process data from human body imaging and standard biobanks. For this purpose, the Minimum Information About Biobank Data Sharing (MIABIS) standard is also used (Scapicchio et al. 2021). An interesting attempt to build a biobank screening system based on the integration of the hospital information system (HIS) and the biobank information system (BIS), which integrated patient identification numbers and was based on natural language processing (NLP) algorithms, was presented by Li et al. (Li et al. 2021). Several other advanced solutions based on AI and ML presently used in population screening may be used to search for potential participants and select interesting digital images and clinical data. An example would be a method of diagnosing diabetic retinopathy using commercial AI-based systems, which have been available on the market since 2018 (Ursin et al. 2021). Another example of the potential of AI in population studies is the use of deep learning algorithms to stratify people at high risk for type 2 diabetes in the general population. These studies are particularly interesting due to the use of a large amount of data from a population biobank. A deep learning model for predicting type 2 diabetes was trained on retinal images from 50,077 UK Biobank donors and evaluated 12,185 study participants. The predictive value of this model in relation to type 2 diabetes was positively assessed (Yun et al. 2022).

Searching for potential participants should respect legal and ethical requirements (Bender et al. 2017). Analyses should be based on the data that users gave consent to process. Any activities that could infringe the privacy of Internet users or social media users should be strictly avoided, and responsible communication based on respect for dignity and autonomy should be pursued.[2] All such activities, both searching via social media or via existing databases, must respect the principle of respect for privacy, ensure data confidentiality, and be based on the informed consent of users.

3 Artificial Intelligence Algorithms in the Consenting Process

The key problem in the functioning of any biobank is obtaining the consent of participants to use their samples and data generated from biological material (Pawlikowski et al. 2011; Bossert et al. 2017). An extremely important issue in this process is the accuracy and comprehensibility of the information provided to potential participants. Some authors report that information provided in the form of an electronic message may have a significant advantage over information provided in the face-to-face form. In this context, AI algorithms and ML may play a significant role in the management of information provided during the process of obtaining informed consent. It may also be particularly useful in the future use of samples and data according to participants' expectations expressed at the time of consent.

AI presents a great opportunity to improve electronic consent in biobanking, particularly dynamic consent. There are many definitions of dynamic consent (e.g., "a personalized communication interface that enables greater participant engagement in clinical and research activities" and "a personalized digital interface that enables two-way communication between participants and researchers") (Kaye et al. 2015), but they all emphasize a participant-centered perspective in decision-making, an interactive IT interface to engage with participants and mutual communication. There are many examples of how dynamic consent is used, such as the First Genetic Trust (2001); the EnCoRe project (Ensuring Consent and Revocation, 2008–2012), where the term "dynamic consent" was used for the first time; InBank; CTRL (digital platform for consent developed by a team from the Australian Genomic Health Alliance); RUDY (a study of rare diseases in the UK); the CHRIS Study (Cooperative Health Research in South Tyrol); and PEER (Platform for Engaging Everyone Responsibly in the USA).

There are many advantages to dynamic consent for both participants and researchers at various stages of research. It enables dynamic (continuous and variable) permission, education, and preferences that improve respect for the ethical principle of autonomy (Dankar et al. 2020). Dynamic permission refers to the inclusion of an online portal to deliver information and allow participants to change permissions/preferences, giving participants the opportunity to accept or decline participation in new studies and to receive individually interpreted results and raw sequence data. Dynamic education relates to the inclusion of a dynamic education component, a dynamic assessment to ensure participant readiness to consent, timely information regarding the study, and the publishing of general and up-to-date information about research progress and outcomes. Dynamic preferences allow participants to select between different levels and types of consent (e.g., broad consent and per-study consent) to tailor the way they receive information, select whether to receive individually interpreted medical results/raw sequence data (and the scope of such results), transfer individual results/information to selected third parties, and opt out of the database at any time. All of these elements can be improved by using AI algorithms to obtain high-quality informed consent, leading to more respect for the principle of autonomy.

Electronic consent presents an interesting opportunity for research ethics. Simon et al. (2022), in a randomized, multicenter study involving 501 patients from two US biobanks, found that electronic consent could lead to better understanding for non-Hispanic patients of higher socioeconomic status. On the other hand, the face-to-face form was associated with a better level of understanding of donors from Hispanic environments and people with lower socioeconomic status. Based on the reports to date, it can be concluded that multimedia can be effective in obtaining consent for biobanking, improving the quality of consent, increasing the amount of time needed for being informed, and possibly even prevailing over face-to-face interviews in some groups (Simon et al. 2016).

It should be said that, so far, the potential of AI, ML, and interactive multimedia to improve the quality of informed consent in population biobanks has not yet been fully explored. Extension of the use of AI algorithms in the process of providing information to participants and obtaining informed consent should be considered. The use of AI algorithms to obtain informed consent should be based on a structured and integrated platform of clinical and biobank data and the interface at the disposal of the donor. Learning algorithms should play an important role in interviewing donors and answering questions. Interactivity is a particularly critical area of AI application. In this context, Karlson et al. (2016) indicate the need to pay attention to the educational content of electronic media, methods of patient authentication in informed consent by electronic means, the need to create a secure IT infrastructure and communication management, and the importance of ethical principles in the process of electronic donor recruitment.

AI algorithms seem to be an interesting tool for better respecting the ethical principle of autonomy by creating an interactive communication platform between biobanks, researchers, and participants. It may improve the transfer of information and communication with research participants at the stage of obtaining informed consent and help develop better solutions for electronic and dynamic consent. It is of significant importance to respect the principle of autonomy, particularly with regard to the future use of samples and data according to participants' will. With the help of AI algorithms, participants can be informed in which research their samples and data are used and can thus control their use, receive feedback on research results, and even object to the use of their samples and data in specific research projects. It may be useful when recontacting participants (when it is needed by biobankers or researchers) as well as in the management of some ethical issues, such as informing about clinically significant incidental findings. It also provides participants with the ability to withdraw consent for the use of their samples and data at any time (which should be an ethical standard in every biomedical research project). However, this tool is limited due to age, education, Internet access, and the digital skills of participants.

4 The Future of AI in Biobanking: A Social "Biofencing" and Humanoid Robots as Research Assistants?

It may be assumed that using AI in population biobanking can play the role of social "biofencing" in the future. By a certain analogy to "geofencing", future (or even current) applications of population biobanking can be forecast. Biobanks might play a role similar to that of GPS in social phenomena management. Geofencing-based systems are intended to position humans and objects in space-time. They manage information about the position and relocation of specifically defined objects. Population biobanks manage the biodata that allows people's "positioning" in the social space. The social and medical phenomenon of premarital genetic screening already used in some parts of the world should be mentioned in this context (Prainsack and Siegal 2006; Altaany et al. 2021; Kisanga et al. 2021; Zeevi et al. 2021). The biodata collected by population biobanks might enable "biocompatible" marriages, where the risk of congenital disorders in offspring is limited as much as possible. Some communities might even go further and implement obligatory premarital genetic screening, which perfectly reflects the biofencing concept by identifying the allowable scope of individuals.

Nonetheless, due to personal data protection systems in Western countries, extensive use of sensitive data is not legally possible. However, it cannot be excluded that the evolution of social consensus might permit it in the future, for example, employers may even be urged to match candidates based on their biological potential (which raises new ethical and legal concerns). In the near future, biofencing may be carried out with the use of expert systems based on AI algorithms. An example of an expert system is the MYCIN expert system created by Edward Shortliffe in 1972, which helped physicians (via dialogue with the physician) select antibiotics (Evans and Pestotnik 1994). A system that synthesizes the capabilities of neural networks with expert systems is known as a knowledge-based artificial neural network (KBANN) (Towell and Shavlik 1994). The result of its operation is the so-called fine-tuning and pre-trained knowledge. By signaling the analyses presented in this chapter, we suggest that biofencing may constitute fine-tuned knowledge in the field of population biobanking.

Expanding the possibilities of using AI in biobanking, it is possible to imagine the integration of AI with robotic systems. Numerous contemporary applications of robotic systems in healthcare offer the prospect, perhaps not far away, of replacing some human roles in medical care and research with humanoid robots. The field of human–robot interaction (HRI) research is multidisciplinary and requires researchers to consider various ethical and psychological issues (Fraune et al. 2022; Dalton-Brown 2020; Massa et al. 2022). Some applications of robotic systems have become a reality in medical care, for example, in the therapy of children with autism spectrum disorders (Mengoni et al. 2017) and in elderly care (Rogers et al. 2022). In a recent meta-analysis carried out by Papadopoulos et al., the authors indicate that the use of humanoid robots in health and social care is limited by technical problems, the capabilities of robots, and negative prejudices regarding the use of robots in healthcare (Papadopoulos 2020). Previous studies have identified some concerns regarding the dissemination of the human–robot relationship in medical practice, such as reliability, safety, and loss of personal care (Broadbent et al. 2010). However, this study was conducted over ten years ago, and social attitudes have evolved with the development of technology toward greater acceptance of human–robot interactions. In our study on a group of 1,000 Polish respondents, we asked about attitudes toward robots in medical care. About 40% of the respondents accepted (12% strongly agree, 28% agree, 20% disagree, and 10% strongly disagree) the possibility of 24-hour care provided by a humanoid robot supervised by means of Internet communication by medical personnel. Similar results were found for consent to be vaccinated against influenza by a humanoid robot (18% strongly agree, 23% agree, 21% disagree, and 18% strongly disagree).[3] In a Eurobarometer survey (2017), 61% of respondents had a positive view of robots and AI, and one-quarter of respondents were comfortable with having a medical operation performed on them by a robot (26%); however, people in different parts of Europe have different ideas of robots (a majority is more likely to think of a machine-like factory robot, but many people also associate it with a human-like robot) (European Commission 2017).

The dynamic development of robotic technology closely related to AI algorithms provokes us to consider the possibility of using humanoid robots in the process of recruiting participants for research and biobanks, particularly in the process of providing information about biobanks and research projects, and collecting samples and data. A humanoid robot could help biobank employees inform and explain biobanking to potential participants. A humanoid robot could also collect biological samples and data. The issue of replacing certain tasks of medical workers with robotic systems has become particularly important in the era of the COVID-19 pandemic because robotic and autonomous systems based on AI are resistant to the virus and are unable to spread pathogens (Shen et al. 2021).

However, ethical and legal issues related to accountability and personal relationships remain problematic. Social research aimed at assessing social attitudes toward robots and the degree of acceptance of such objects should be conducted.[4] Therefore, the use of robots cannot be viewed as a replacement for humans; instead, it acts as an aid in carrying out certain tasks. In addition, the responsibility must always be assigned to the human who supervises the robot's work, that is, a researcher, physician, biobanker, nurse, diagnostician, or any other authorized person.

5 Conclusion

Technologies based on AI, despite many ethical and legal concerns, could be a breakthrough tool in improving communication with participants in the biobanking of HBM and, more broadly, in biomedical research and clinical care. AI algorithms can be used to search for and enroll potential participants in biobanks based on their demographic, social, psychological, and even medical characteristics, as well as to conduct comprehensive analyses based on linking research databases with medical records. It may improve the level of respect for the fundamental rights of research participants, particularly autonomy, through its interactive platform, mutual communication, and participant-centered decision-making. It may help develop better solutions for electronic and dynamic consent and manage some ethical issues, such as future use, recontacting participants, consent withdrawal, and informing about clinically significant findings. Responsible implementation of new technologies may contribute to increasing respect for the autonomy and dignity of participants and to improving the level of social trust in biobanks and medical research. However, the use of AI requires special protection of other human rights, such as privacy, which should be expressed by data confidentiality and a responsible data sharing and management policy. Other ethical and legal issues connected with AI in biobanking related to accountability, responsible government, access, and the sharing of samples and data should be considered. Finally, the protection of personal relationships in the context of the dissemination of innovative technologies based on AI remains a challenge and requires deep ethical sensitivity and psychological reflection.

Notes

* The authors are affiliated to BBMRI.pl, the Biobanking and BioMolecular resources Research Infrastructure Poland as ELSI experts.

1 For more on potentiality, see chapter, "Powerful AI: Between Accountability and Potentiality" by Michaela Th. Mayrhofer.

2 For more on communication and participatory approaches, see chapter "Tackling AI Transparency Concerns in Biomedical Research: Bringing a Communication-Participatory Approach to the Conversation" by Leslie Salgado Arzuaga.

3 The study titled "Social attitudes towards artificial intelligence and humanoid robots in medical care" was carried out by Jaroslaw Sak and Jakub Pawlikowski and is unpublished. The first question was: "If it was necessary to provide 24-hour care, would you – if there was such a need for health reasons – consent to such care carried out around the clock only by a humanoid robot (in appearance similar to a human) supervised via internet links by medical personnel?". The second question was: "Would you give consent for a humanoid robot to vaccinate you against influenza instead of a nurse?". The study was carried out in 2019 on 1,000 Polish adults.

4 For further insights on social acceptance, trust, and AI, see chapters "Social Acceptance of Artificial Intelligence in Biobanking" by Elodie Caboux, Patrick Courtney, Io Hong Cheong, and Zisis Kozlakidis and "Unique by Design–The Need for Individualized Models of Trust in Biobanking AI" by Anna Clareborn.

Reference List

Abdullah, Y. I., J. S. Schuman, R. Shabsigh, A. Caplan, and L. A. Al-Aswad. 2021. "Ethics of Artificial Intelligence in Medicine and Ophthalmology." *Asia-Pacific Journal of Ophthalmology (Philadelphia, PA)* 10 (3): 289–98. https://doi.org/10.1097/apo.0000000000000397.

Abu Farha, R., K. H. Alzoubi, O. F. Khabour, and T. L. Mukattash. 2020. "Factors Influencing Public Knowledge and Willingness to Participate in Biomedical Research in Jordan: A National Survey." *Patient Prefer Adherence* 14: 1373–9. https://doi.org/10.2147/ppa.S261903.

Ahram, Mamoun, Areej Othman, Manal Shahrouri, and Ebtihal Mustafa. 2014. "Factors Influencing Public Participation in Biobanking." *European Journal of Human Genetics* 22 (4): 445–51. https://doi.org/10.1038/ejhg.2013.174.

Altaany, Z., O. F. Khabour, K. H. Alzoubi, A. K. Alkaraki, and G. Al-Taani. 2021. "The Perception of Premarital Genetic Screening within Young Jordanian Individuals." *Public Health Genomics* 24 (3–4): 182–8. https://doi.org/10.1159/000517162.

Bender, Jacqueline Lorene, Alaina B. Cyr, Luk Arbuckle, and Lorraine E. Ferris. 2017. "Ethics and Privacy Implications of Using the Internet and Social Media to Recruit Participants for Health Research: A Privacy-by-Design Framework for Online Recruitment." *Journal of Medical Internet Research* 19 (4): e104. https://doi.org/10.2196/jmir.7029.

Bossert, Sabine, Hannes Kahrass, Ulrike Heinemeyer, Jana Prokein, and Daniel Strech. 2017. "Participatory Improvement of a Template for Informed Consent Documents in Biobank Research – Study Results and Methodological Reflections." *BMC Medical Ethics* 18 (1): 78. https://doi.org/10.1186/s12910-017-0232-7.

Brailean, A., J. Curtis, K. Davis, A. Dregan, and M. Hotopf. 2020. "Characteristics, Comorbidities, and Correlates of Atypical Depression: Evidence from the UK Biobank Mental Health Survey." *Psychological Medicine* 50 (7): 1129–38. https://doi.org/10.1017/s0033291719001004.

Broadbent, E., I. H. Kuo, Y. I. Lee, J. Rabindran, N. Kerse, R. Stafford, and B. A. MacDonald. 2010. "Attitudes and Reactions to a Healthcare Robot." *Telemedicine Journal and E-Health* 16 (5): 608–13. https://doi.org/10.1089/tmj.2009.0171.

Broekstra, R., J. Aris-Meijer, E. Maeckelberghe, R. Stolk, and S. Otten. 2021. "Demographic and Prosocial Intrapersonal Characteristics of Biobank Participants and Refusers: The Findings of a Survey in the Netherlands." *European Journal of Human Genetics* 29 (1): 11–19. https://doi.org/10.1038/s41431-020-0701-1.

Cahill, Tiernan J., Blake Wertz, Qiankun Zhong, et al. 2019. "The Search for Consumers of Web-Based Raw DNA Interpretation Services: Using Social Media to Target Hard-to-Reach Populations." *Journal of Medical Internet Research* 21 (7): e12980. https://doi.org/10.2196/12980.

Chen, Chen, Wenjia Bai, Rhodri H. Davies, et al. 2020. "Improving the Generalizability of Convolutional Neural Network-Based Segmentation on CMR Images." *Frontiers in Cardiovascular Medicine* 7. https://doi.org/10.3389/fcvm.2020.00105.

Dalton-Brown, S. 2020. "The Ethics of Medical AI and the Physician-Patient Relationship." *Cambridge Quarterly of Healthcare Ethics* 29 (1): 115–21. https://doi.org/10.1017/s0963180119000847.

Dankar, Fida K., Marton Gergely, Bradley Malin, Radja Badji, Samar K. Dankar, and Khaled Shuaib. 2020. "Dynamic-Informed Consent: A Potential Solution for Ethical Dilemmas in Population Sequencing Initiatives." *Computational and Structural Biotechnology Journal* 18: 913–21. https://doi.org/10.1016/j.csbj.2020.03.027.

Darko, E. M., M. Kleib, and J. Olson. 2022. "Social Media Use for Research Participant Recruitment: Integrative Literature Review." *Journal of Medical Internet Research* 24 (8): e38015. https://doi.org/10.2196/38015.

Domaradzki, Jan, and Jakub Pawlikowski. 2019. "Public Attitudes toward Biobanking of Human Biological Material for Research Purposes: A Literature Review." *International Journal of Environmental Research and Public Health* 16 (12). https://doi.org/10.3390/ijerph16122209.

Escott-Price, V., D. J. Smith, K. Kendall, et al. 2019. "Polygenic Risk for Schizophrenia and Season of Birth Within the UK Biobank Cohort." *Psychological Medicine* 49 (15): 2499–504. https://doi.org/10.1017/s0033291718000454.

European Commission. Directorate-General for Communication. 2017. "Special Eurobarometer 460: Attitudes Towards the Impact of Digitization and Automation on Daily Life" v 1.00.

Evans, R. Scott, and Stanley L. Pestotnik. 1994. "Applications of Medical Informatics in Antibiotic Therapy." In *Antimicrobial Susceptibility Testing: Critical Issues for the 90s*, edited by James A. Poupard, Lori R. Walsh and Bruce Kleger, 87–96. Boston, MA: Springer US.

Fotaki, Anastasia, Esther Puyol-Antón, Amedeo Chiribiri, René Botnar, Kuberan Pushparajah, and Claudia Prieto. 2022. "Artificial Intelligence in Cardiac MRI: Is Clinical Adoption Forthcoming?" *Frontiers in Cardiovascular Medicine* 8. https://doi.org/10.3389/fcvm.2021.818765.

Fraune, Marlena R., Iolanda Leite, Nihan Karatas, et al. 2022. "Lessons Learned About Designing and Conducting Studies from HRI Experts." *Frontiers in Robotics and AI* 8. https://doi.org/10.3389/frobt.2021.772141.

Gaskell, G., H. Gottweis, J. Starkbaum, et al. 2013. "Publics and Biobanks: Pan-European Diversity and the Challenge of Responsible Innovation." *European Journal of Human Genetics* 21 (1): 14–20. https://doi.org/10.1038/ejhg.2012.104.

Gerke, Sara, Timo Minssen, and Glenn Cohen. 2020. "Ethical and Legal Challenges of Artificial Intelligence-Driven Healthcare." *Artificial Intelligence in Healthcare*. https://doi.org/10.1016/B978-0-12-818438-7.00012-5.

Han, X., K. Steven, A. Qassim, et al. 2021. "Automated AI Labeling of Optic Nerve Head Enables Insights into Cross-Ancestry Glaucoma Risk and Genetic Discovery in >280,000 Images from UKB and CLSA." *American Journal of Human Genetics* 108 (7): 1204–16. https://doi.org/10.1016/j.ajhg.2021.05.005.

Hunter, R. F., A. Gough, N. O'Kane, et al. 2018. "Ethical Issues in Social Media Research for Public Health." *American Journal of Public Health* 108 (3): 343–8. https://doi.org/10.2105/ajph.2017.304249.

Kaneko, Masatomo, Norio Fukuda, Hitomi Nagano, et al. 2022. "Artificial Intelligence Trained with Integration of Multiparametric MR-US Imaging Data and Fusion Biopsy Trajectory-Proven Pathology Data for 3D Prediction of Prostate Cancer: A Proof-of-Concept Study." *The Prostate* 82 (7): 793–803. https://doi.org/10.1002/pros.24321.

Karlson, E. W., N. T. Boutin, A. G. Hoffnagle, and N. L. Allen. 2016. "Building the Partners HealthCare Biobank at Partners Personalized Medicine: Informed Consent, Return of Research Results, Recruitment Lessons and Operational Considerations." *Journal of Personalized Medicine* 6 (1). https://doi.org/10.3390/jpm6010002.

Kaye, Jane, Edgar A. Whitley, David Lund, Michael Morrison, Harriet Teare, and Karen Melham. 2015. "Dynamic Consent: A Patient Interface for Twenty-First Century Research Networks." *European Journal of Human Genetics* 23 (2): 141–6. https://doi.org/10.1038/ejhg.2014.71.

Kinkorová, Judita, and Ondřej Topolčan. 2020. "Biobanks in the Era of Big Data: Objectives, Challenges, Perspectives, and Innovations for Predictive, Preventive, and Personalised Medicine." *EPMA Journal* 11 (3): 333–41. https://doi.org/10.1007/s13167-020-00213-2.

Kisanga, E., R. Mutagonda, D. T. Myemba, et al. 2021. "Premarital Genetic Screening and Care of Tanzanian Children with Sickle Cell Disease: A Qualitative Study on Parents' Views and Experiences." *Journal of Community Genetics* 12 (4): 515–23. https://doi.org/10.1007/s12687-021-00539-y.

Klassen, Karen M., Caitlin H. Douglass, Linda Brennan, Helen Truby, and Megan S. C. Lim. 2018. "Social Media Use for Nutrition Outcomes in Young Adults: A Mixed-methods Systematic Review." *International Journal of Behavioral Nutrition and Physical Activity* 15 (1): 70. https://doi.org/10.1186/s12966-018-0696-y.

Kolluri, Sheela, Jianchang Lin, Rachael Liu, Yanwei Zhang, and Wenwen Zhang. 2022. "Machine Learning and Artificial Intelligence in Pharmaceutical Research and Development: A Review." *The AAPS Journal* 24 (1): 19. https://doi.org/10.1208/s12248-021-00644-3.

Kozlakidis, Zisis. 2020. "Biobanks and Biobank-Based Artificial Intelligence (AI) Implementation Through an International Lens." In *Artificial Intelligence and Machine Learning for Digital Pathology, State-of-the-Art and Future Challenges*, edited by Andreas Holzinger, Randy Goebel, Michael Mengel and Heimo Müller. Springer International Publishing.

Li, X., J. Han, S. Zhang, K. Chen, L. Zhao, Y. He, and S. Liu. 2021. "Artificial Intelligence for Screening Chinese Electronic Medical Record and Biobank Information." *Biopreserv Biobank* 19 (5): 386–93. https://doi.org/10.1089/bio.2020.0151.

Manyika, James, Michael Chui, Brad Brown, et al. 2011. "Big Data: The Next Frontier for Innovation, Competition, and Productivity." *McKinsey*. www.mckinsey.com/business-functions/business-technology/our-insights/big-data-the-next-frontier-for-innovation.

Massa, N., P. Bisconti, and D. Nardi. 2022. "The Psychological Implications of Companion Robots: A Theoretical Framework and an Experimental Setup." *International Journal of Social Robotics*: 1–14. https://doi.org/10.1007/s12369-021-00846-x.

McCradden, Melissa D., James A. Anderson, Elizabeth A. Stephenson, et al. 2022. "A Research Ethics Framework for the Clinical Translation of Healthcare Machine Learning." *The American Journal of Bioethics* 22 (5): 8–22. https://doi.org/10.1080/15265161.2021.2013977.

Mengoni, S. E., K. Irvine, D. Thakur, et al. 2017. "Feasibility Study of a Randomised Controlled Trial to Investigate the Effectiveness of Using a Humanoid Robot to Improve the Social Skills of Children with Autism Spectrum Disorder (Kaspar RCT): A Study Protocol." *BMJ Open* 7 (6): e017376. https://doi.org/10.1136/bmjopen-2017-017376.

Müller, Vincent C. 2021. "Ethics of Artificial Intelligence and Robotics." In *The Stanford Encyclopedia of Philosophy Summer 2021*, edited by Edward N. Zalta and Uri Nodelman. Stanford University.

Murphy, Kathleen, Erica Di Ruggiero, Ross Upshur, et al. 2021. "Artificial Intelligence for Good Health: A Scoping Review of the Ethics Literature." *BMC Medical Ethics* 22 (1): 14. https://doi.org/10.1186/s12910-021-00577-8.

Narita, Akira, Masao Ueki, and Gen Tamiya. 2021. "Artificial Intelligence Powered Statistical Genetics in Biobanks." *Journal of Human Genetics* 66 (1): 61–5. https://doi.org/10.1038/s10038-020-0822-y.

Papadopoulos, I., C. Koulouglioti, R. Lazzarino, and S. Ali. 2020. "Enablers and Barriers to the Implementation of Socially Assistive Humanoid Robots in Health and Social Care: A Systematic Review." *BMJ Open* 10 (1): e033096. https://doi.org/10.1136/bmjopen-2019-033096.

Pawlikowski, Jakub, Jaroslaw Sak, and Krzysztof Marczewski. 2011. "Biobank Research and Ethics: The Problem of Informed Consent in Polish Biobanks." *Archives of Medical Science* 7 (5): 896–901.

Pawlikowski, Jakub, Michał Wiechetek, and Anita Majchrowska. 2022. "Associations between the Willingness to Donate Samples to Biobanks and Selected Psychological Variables." *International Journal of Environmental Research and Public Health* 19 (5): 2552. https://doi.org/10.3390/ijerph19052552.

Polimanti, R., R. E. Peterson, J. S. Ong, et al. 2019. "Evidence of Causal Effect of Major Depression on Alcohol Dependence: Findings from the Psychiatric Genomics Consortium." *Psychological Medicine* 49 (7): 1218–26. https://doi.org/10.1017/s0033291719000667.

Prainsack, Barbara, and Gil Siegal. 2006. "The Rise of Genetic Couplehood? A Comparative View of Premarital Genetic Testing." *BioSocieties* 1 (1): 17–36. https://doi.org/10.1017/S1745855205050106.

Prictor, M., M. A. Lewis, A. J. Newson, et al. 2020. "Dynamic Consent: An Evaluation and Reporting Framework." *Journal of Empirical Research on Human Research Ethics* 15 (3): 175–86. https://doi.org/10.1177/1556264619887073.

PRIMAGE. n.d. "Medical Imaging, Artificial Intelligence, Childhood Cancer Research." 2023. www.primageproject.eu/ (20.06.2023).

Rahm, Alanna Kulchak, Michelle Wrenn, Nikki M. Carroll, and Heather Spencer Feigelson. 2013. "Biobanking for Research: A Survey of Patient Population Attitudes and Understanding." *Journal of Community Genetics* 4 (4): 445–50. https://doi.org/10.1007/s12687-013-0146-0.

Realo, A., H. van Middendorp, L. Kööts-Ausmees, J. Allik, and A. W. M. Evers. 2018. "Role of Personality Traits in Reporting the Development of Adverse Drug Reactions: A Prospective Cohort Study of the Estonian General Population." *BMJ Open* 8 (7): e022428. https://doi.org/10.1136/bmjopen-2018-022428.

Rogers, W. A., T. Kadylak, and M. A. Bayles. 2022. "Maximizing the Benefits of Participatory Design for Human-Robot Interaction Research with Older Adults." *Human Factors* 64 (3): 441–50. https://doi.org/10.1177/00187208211037465.

Ronaldson, A., J. Arias de la Torre, F. Gaughran, I. Bakolis, S. L. Hatch, M. Hotopf, and A. Dregan. 2022. "Prospective Associations between Vitamin D and Depression in Middle-Aged Adults: Findings from the UK Biobank Cohort." *Psychological Medicine* 52 (10): 1866–74. https://doi.org/10.1017/s0033291720003657.

Sallis, H. M., G. Davey Smith, and M. R. Munafò. 2019. "Cigarette Smoking and Personality: Interrogating Causality Using Mendelian Randomisation." *Psychological Medicine* 49 (13): 2197–205. https://doi.org/10.1017/s0033291718003069.

Scapicchio, Camilla, Michela Gabelloni, Sara Maria Forte, et al. 2021. "DICOM-MIABIS Integration Model for Biobanks: A Use Case of the EU PRIMAGE Project." *European Radiology Experimental* 5 (1): 20. https://doi.org/10.1186/s41747-021-00214-4.

Shen, Y., D. Guo, F. Long, et al. 2021. "Robots Under COVID-19 Pandemic: A Comprehensive Survey." *IEEE Access* 9: 1590–615. https://doi.org/10.1109/ACCESS.2020.3045792.

Simon, C. M., D. W. Klein, and H. A. Schartz. 2016. "Interactive Multimedia Consent for Biobanking: A Randomized Trial." *Genetics in Medicine* 18 (1): 57–64. https://doi.org/10.1038/gim.2015.33.

Simon, C. M., K. Wang, L. A. Shinkunas, et al. 2022. "Communicating with Diverse Patients About Participating in a Biobank: A Randomized Multisite Study Comparing Electronic and Face-to-Face Informed Consent Processes." *Journal of Empirical Research on Human Research Ethics* 17 (1–2): 144–66. https://doi.org/10.1177/15562646211038819.

Towell, Geoffrey G., and Jude W. Shavlik. 1994. "Knowledge-Based Artificial Neural Networks." *Artificial Intelligence* 70 (1): 119–65. https://doi.org/10.1016/0004-3702(94)90105-8.

Urias, Müller G., Niravkumar Patel, Changyan He, et al. 2019. "Artificial Intelligence, Robotics and Eye Surgery: Are We Overfitted?" *International Journal of Retina and Vitreous* 5 (1): 52. https://doi.org/10.1186/s40942-019-0202-y.

Ursin, Frank, Cristian Timmermann, Marcin Orzechowski, and Florian Steger. 2021. "Diagnosing Diabetic Retinopathy with Artificial Intelligence: What Information Should Be Included to Ensure Ethical Informed Consent?" *Frontiers in Medicine* 8. https://doi.org/10.3389/fmed.2021.695217.

Vijayan, R. S. K., Jan Kihlberg, Jason B. Cross, and Vasanthanathan Poongavanam. 2022. "Enhancing Preclinical Drug Discovery with Artificial Intelligence." *Drug Discovery Today* 27 (4): 967–84. https://doi.org/10.1016/j.drudis.2021.11.023.

Whitaker, C., S. Stevelink, and N. Fear. 2017. "The Use of Facebook in Recruiting Participants for Health Research Purposes: A Systematic Review." *Journal of Medical Internet Research* 19 (8): e290. https://doi.org/10.2196/jmir.7071.

Yun, J. S., J. Kim, S. H. Jung, et al. 2022. "A Deep Learning Model for Screening Type 2 Diabetes from Retinal Photographs." *Nutrition, Metabolism & Cardiovascular Diseases* 32 (5): 1218–26. https://doi.org/10.1016/j.numecd.2022.01.010.

Zeevi, D. A., W. K. Chung, C. Levi, et al. 2021. "Recommendation of Premarital Genetic Screening in the Syrian Jewish Community Based on Mutation Carrier Frequencies within Syrian Jewish Cohorts." *Molecular Genetics & Genomic Medicine* 9 (8): e1756. https://doi.org/10.1002/mgg3.1756.

8 Tackling AI Transparency Concerns in Biomedical Research

Bringing a Communication-Participatory Approach to the Conversation

Leslie Salgado Arzuaga

1 Introduction

In a discourse dominated by promises of enhanced efficiency (Cadario et al. 2021), the tech industry now recognizes "that a lack of public trust may pose one of the greatest barriers to AI adoption in areas ranging from driverless cars to healthcare" (Garvey and Maskal 2020, 287). Although the studies of trust in AI are still an emergent area, there seems to be some consensus that some systems' black box nature "can lead to a lack of trust (worthiness) in and acceptance of AI systems by clinicians and patients" (Goisauf and Abadía 2022, 8). In biomedical research, particularly biobanking, empirical evidence shows how the lack of trust can hinder public support and participation, "which may translate into activities being stopped or abandoned" (O'Doherty et al. 2011, 369). It is clear, then, that the challenges that the adoption of AI-powered technologies brings to the field, concerning lack of public trust, cannot be overlooked. Due to the difficulties of understanding how some AI systems make decisions and the need to make people more knowledgeable about AI (Cadario et al. 2021; Goisauf and Abadía 2022), "transparency" is a critical principle to follow in building public trust in AI-driven technologies. Indeed, it has been identified as the most common and one of the fundamental five principles in many ethical guidelines addressing AI globally (Jobin et al. 2019). Among various requirements for trustworthiness AI, the AI High-Level Expert Group (HLEG) "Ethics Guidelines for Trustworthy Artificial Intelligence" (2019) connects transparency with traceability, explainability, and clear communication in the AI creation and adoption process.

Building upon that approach, the European Union AI Act (European Commission 2021) adopts a tiered, risk-based approach to transparency. For example, for systems in the "high-risk" tier, "transparency" is a central aspect, with the Act mandating that "to address the opacity that may make certain AI systems incomprehensible to or too complex for natural persons, a certain degree of transparency should be required for high-risk AI systems" (European Commission 2021, recital 47, p. 30). Adding that "users should be able to interpret the system output and use it appropriately (Idem)", which aligns with a perspective focused on the links between transparency and public trust (EU 2020).

In biobanking, "transparency" has been recognized as one of the future-proofing biobanks' governance principles linked to sharing and disclosing information,

DOI: 10.4324/9781032634487-12

namely, the "availability of information about an actor allowing other actors to monitor the working or performance of this actor" (Gille et al. 2020, 989). Ethical governance is associated with transparency, accountability, and oversight in the biobank literature.[1] But what happens when those actors are AI-powered systems whose decisions are difficult to explain and monitor, even for their creators? Investigation suggests that the "future-proofing biobanks' governance principles" "can also be applied to AI ethics in biomedical research institutions" (Kargl et al. 2022, 157), including "transparency" as information sharing and openness. However, applying transparency is far from being straightforward. The unexplainable nature of some AI models and the need to better understand transparency's social dimension in the AI context lay behind some of the difficulties of putting this principle into practice. To contribute to advancing this conversation, this chapter proposes a dialogue between the literature on the social dimensions of AI transparency, explainability, and interpretability and learnings from studies concerning public engagement in biobanking. Developing a participatory and communicational understanding of transparency can help us apply this principle while adopting AI-driven technologies in biomedical research.

To contextualize this proposal, the chapter will bring scholarly discussion concerned with transparency, explainability, and interpretability from a social sciences perspective. Then, it will analyze two cases of study against this backdrop and through the lenses of socio-constructivist theories of technology. By analyzing the cases of the AI-powered applications "HistoMapr™" (currently PredxBio Platform)[2] and "Explainable AI in Dermatology (ExAID)", the chapter will illustrate two approaches to embracing communicative and participatory viewpoints on AI transparency and explainability in biomedical research. The cases were purposively selected because they represent distinct approaches to the same phenomenon and are anchored into the biomedical field. Finally, it will discuss how a broader approach to transparency and explainability can help us to (1) understand transparency as a feature of the innovation and adoption processes, (2) embrace co-design strategies to enhance transparency, and (3) incorporate built-in and external solutions to provide interpretable explanations.

1.1 AI Explainability, Transparency, and Interpretability – a Look at Some Social Perspectives

"Transparency", "explainability", and "interpretability" often appear as interconnected concepts in AI literature.[3] The World Health Organization (WHO 2021) Guidance on Ethics of AI delineates transparency and explainability as principles necessary for the intelligibility of AI systems. Reflecting on AI's lack of transparency as a key challenge in biobanking, the literature emphasizes the intertwined nature of these ethical principles: "Transparency relates most importantly to model interpretability or the 'Explainable AI' notion – in other words, humans should be able to understand or interpret how a given technology reached a certain decision or prediction" (Kozlakidis 2020, 200).

This approach aligns with transparency as the opposite of "opacity or black box-ness" (Lipton 2017, 4). Widely accepted, this standpoint relates to the ontological dimension of transparency in computer science, which focuses on whether AI models' decisions are explainable to humans. In line with this stance, Lipton (2017) identifies three levels of AI model transparency: (1) at the entire model level (simulatability), (2) at the level of individual components (decomposability), and (3) at the level of the training algorithm (algorithmic transparency). Slightly broader but still in a similar breath, Matheny et al. (2019) propose separating transparency into three categories: (1) data transparency, (2) algorithmic transparency, and (3) performance reporting.

However, even if we assume "transparency" as an achievable end, reaching "transparent" models is challenging due to their inherent complexity – particularly neural networks, which remain unintelligible even to their creators (Christian 2021). Acknowledging that some complex models may not always be explainable (Rudin 2019) is essential for biomedical research, as it may influence the selection of interpretable[4] versus larger, unexplainable models when developing AI-powered technologies.

To address challenges associated with the black box nature of some systems, computer scientists have developed post hoc explainability methods, including "analysis of feature importance, saliency maps, counterfactual methods, recognizing textual entailment, and knowledge distillation" (Gyevnar et al. 2023, 965). However, exposing these methods to a communication-anchored analysis highlights their typical lack of intelligibility. For example, methods such as SHAP[5] and LIME[6] produce explanations that may be difficult for nonspecialists to interpret. Despite advancements in this area, the challenge remains: making explainability methods "accessible, comprehensible, and useful" (Mittelstadt et al. 2019, 281). This underscores the importance of adopting a multidisciplinary perspective to transparency and explainability, where a communicative approach is critical. It is about remembering that "the symbols and metaphors used for communication, that is, mathematically founded algorithms, may be dependent on translations to human imaginaries and languages" (Larsson and Heintz 2020, 8). This discussion highlights two critical issues: the need to embrace "transparency" in its profound implications for user trust and broadening "explainability" beyond technical solutions.

Adopting transparency and explainability in their socio-technical nature is crucial in light of recent legal developments, particularly the EU AI Act. While the Act adopts a broad socio-technical notion of transparency, concerns about a "transparency gap" have arisen (Gyevnar et al. 2023). This term refers to the divergence between the EU AI Act's holistic transparency framework and explainable AI (xAI) narrower focus, prioritizing algorithmic transparency. These authors suggest bridging this gap requires xAI to "develop new approaches that support legal frameworks in achieving transparency objectives" (Gyevnar et al. 2023, 965). The "transparency gap" debate reinforces the imperative of integrating insights from the humanities and social sciences into this (Larsson and Heintz 2020; Mittelstadt et al. 2019). In a socio-technical approach, incorporating communicative-participatory perspectives to transparency and explainability could enhance the interpretability of salience methods and contribute to filling the gap.

A key lesson from examining humanities and social sciences frameworks is that "transparency is performed by socio-technical assemblages of algorithm and people" (Kemper and Kolkman 2019, 2083). From this idea emerges the need to contextualize potential solutions to address transparency in a use-case differential fashion (Matheny et al. 2019). This implies considering the arrangements of the particular situation where the AI-driven solution will be deployed. For example, pursuing transparency in adopting AI-powered technology for a biobank network (Dive et al. 2020) would significantly differ from designing solutions for one entity to another. If we consider issues like data curation or the robustness of the data security system, the challenges and solutions will vary from a networked context to a "single" biobank context.

Thinking of transparency's role in the ethical governance of biobanks, for instance, it is straightforward that, uniquely, a computational approach does not fulfill its role in biomedical research or its multifaceted character (Oxelheim and Forssbæck 2014).

Under the "Necessary Conditions for Trustworthy Biobank Governance", researchers remark transparency as an "overview of operations, and decision-making is open to scrutiny" (O'Doherty et al. 2011, 369). As it is noticeable, this approach is mainly related to the socio-technical and normative ideal of "openness", which intersects with recent proposals to approach AI transparency from a social sciences perspective. Like in the biobanking field (Gille et al. 2020; Kargl et al. 2022), transparency has been linked to public trust in the governance structures in social sciences, law, and humanities studies.[7]

1.2 AI Transparency in Its Communicative Dimension

Approaches to AI transparency with communicative-participatory breath can take different forms. The transparency as design publicity approach – for example – unpacks different stances of transparency, namely: "(1) deploying an explanation 'by design'; (2) an intentional explanation of its deployment; (3) a justification of its use; and (4) when used consistently, a procedural justification of the individual decisions it takes" (Loi et al. 2021, 262). While this proposal is focused on the functional stances, approaches according to the shareholder's role in AI development have also emerged (Fritzsche et al. 2023; Matheny et al. 2019).

From a communication standpoint, it is essential to think about transparency as a framework to reduce asymmetries in access to information (Oxelheim and Forssbæck 2014). In the context of AI in biobanking, providing sufficient information about data security mechanisms, the paths of data curation processes, and the rationale for choosing a large or simpler model, for example, can influence the decisions and inputs of stakeholders at different levels and with various roles in the innovation process. However, the key to making "transparency" meaningful from a communication perspective is "making the information . . . not only observable but also relevant to the receiver, trustworthy, and sufficiently precise" (Oxelheim and

Forssbæck 2014, 35). Following this principle, embracing transparency in developing and adopting AI systems means making them (and their decisions) explainable in a manner adapted to the stakeholders concerned (WHO 2021). The distinction between xAI, which mainly refers to an "explanatory agent revealing underlying causes to it or another agent's decision making" (Miller 2019, 2), and "everyday explanations"[8] can help us to design strategies to develop intelligible communication for AI transparency, considering the differentiated informational needs of different stakeholders. This approach reminds us that explanations can serve different functions, like complaining about existing legislation or explaining the model performance, and different forms, like the inclusion of a "why did you do that" button (Loi et al. 2021) in AI interacting with humans or communicative external products. While the approaches might vary widely, lessons from cognitive sciences reveal critical elements to consider.

After surveying 250 documents focused on AI "explanations", Miller (2019) reminded us about key features: contrastive, selected (in a biased manner), social, and causal (pointing out that probabilities are not as crucial as causal links). These ideas can be incorporated into the development of salience methods and into the design of communicational methods added or external to the model.

Acknowledging that "a large portion of this [transparency-explainability-interpretability] – still unexplored is thinking about the human side" (Christian 2021, 113) also means recognizing the agency of the actors in the innovation process, allowing us to focus on creating the conditions to achieve "interpretability" and the purposes of explanations, including the data, the performance, and the model outcomes. Although the borders between the three notions are frequently diffused in the literature, "interpretability" often refers to "how well one can understand how an AI system works, while explainability refers to how well one can explain what happens in AI decision-making in understandable terms" (Goisauf and Abadía 2022, 7). It is about transitioning from providing explanations to providing conditions for interpretability, highlighting the user's agency in the assemblage of innovation.

The specific case of AI-driven polygenic scores (PRSs) reminds us that "the risks of misinterpretation of (AI-driven) PRSs by clinicians, patients, and other stakeholders involved should not be underestimated, especially as there may be a risk of drawing conclusions about causal relationships too quickly" (Slunecka et al. 2021 in Fritzsche et al. 2023, 3). This approach underscores the diversity of factors influencing interpretations, including geography, socioeconomic conditions, knowledge, and cultural background of the technology's "relevant social groups".

The review of the social sciences literature preoccupied with AI transparency, explainability, and interpretability reveals a growing interest in bringing decades of learning in these fields into AI technology development. There is a common concern about embracing transparency throughout AI innovation and adoption. The value attributed to intelligible, clear, situated information sharing is evident, considering differentiated needs, actors, and situations during the whole AI life cycle. The case studies in the next section bring practical insights to real-life scenarios where a situated and socio-technical understanding of explainability-interpretability enhances the systems' transparency.

2 Addressing Transparency and Explainability: Two Approaches in Biomedical Research

This section explores two cases that illustrate different approaches for addressing some of the transparency challenges AI technologies entail for biomedical research and medicine. While the AI (xAI) Platform for Computational Pathology Solutions HistoMapr™ is a by-design solution to address explainability, the Deep Learning Based Skin Lesion Classifier follows a co-design approach to make the tool interpretable to different stakeholders. Although the approaches differ, both cases follow the notion of transparency as synonymous with openness and access to intelligible information.

2.1 Case 1: HistoMaprTM

According to the information published on the PredxBio (2023) website, HistoMapr™ is a proprietary software platform created by this company that can predict tumor initiation, progression, and response to therapy using explainable AI and unbiased spatial analytics. It recently was rebranded as PredxBio Platform. The system analyzes Whole Slides Images (WSI) of tissues. It uses an interface centered on the pathologist to show the Region of Interest (ROI) in the triage order so that the pathologist can see malignant or atypical areas that might be present. The explainability feature is expressed through a "why" button. Using the pathologist's language (e.g., robust and rigid architecture, and highly monomorphic nuclear patterns), the interface shows an explanatory information field about why the case is triaged this way and whether the pathologist can agree or not if some doubts emerge. This solution "fleshes out the relevant differential diagnosis and acknowledges the strength of HistoMapr's analysis via a Confidence Score" (Tosun et al. 2020, 217).

The responses provided by the "why" button and the pathologist's role during the process are critical features of this tool that align with recommendations based on research on explainability and transparency in philosophy, cognitive science, and the social sciences. This approach is coherent with proposals emerging from cognitive and social sciences discussed earlier,[9] which suggest that why-questions require contrastive, selective, and socially interactive explanations. The contrastive dimension is given by the fact that the explanations identify and flesh out the relevant differential diagnosis, providing supplementary data for the professional to compare and interpret relevant information.

Using the pathologist's language enhances the "interpretable" character of the information, which improves the chances for the professional to produce clear and intelligible explanations of the results to other actors, for example, patients and physicians. Interpretable information has the potential to facilitate the informed dialogue among actors (Mittelstadt et al. 2019), which constitutes another element identified in the literature to broaden our understanding of what constitutes transparency and explainability in the innovation and adoption of AI.

This tool aims to augment rather than to substitute the pathologist's role (Tosun et al. 2020). It also addresses the "agency" issue, which is essential to this debate because, by incorporating a situated approach to explainability, it embraces a non-determinist framework to AI adoption.

Philosophically defined as "standards of judgment and fosters attitudes which make men ready to accept and even to introcept the dictates of the apparatus" (Marcuse 1982, 41),[10] a "formal rationality" seems to prevail in AI development (Webber and Prietl 2022), which influences a framework of adoption where human judgment is often replaced by the machine's outcome.[11] Acknowledging the user's agency in interacting with AI-driven tools entails embracing an alternative paradigm. The fact that HistoMapr™ offers pathologists the opportunity to express agreement, disagreement, or indicate uncertainty, collecting this information for possible additional stain workup or consultation (Tosun et al. 2020) illustrates how a user-centered approach to explainability acknowledges user agency by design.

2.2 Case 2: Deep Learning-Based Skin Lesion Classifier

The case "Explainable AI in Dermatology" (ExAID), created at the German Research Center for Artificial Intelligence (DFKI, by German acronyms) (Lucieri et al. 2020), will serve to illustrate a co-design process aligned with a participatory approach to AI development. In the early design phase of this AI system, the team embraced a co-design strategy following the Z-inspection® (Zicari et al. 2021a) framework, intended to "satisfy the requirement for Trustworthy AI as defined by the High-Level Expert Group on AI (AI HLEG) set up by the European Commission" (Zicari et al. 2021b).

The lack of explainability, and consequently of acceptability of the Computer-Assisted Diagnostic systems in diagnosing malignant melanoma, motivated the team to create a system to provide dermatologists with interpretable explanations that can support the diagnostic process (Lucieri et al. 2020). In response to those challenges, ExAID provides "intuitive intelligibility for end users by conveying multi-modal decision justification centered around standard concepts commonly used in the dermatology domain" (Lucieri et al. 2022, 2). This was translated into post hoc explanations in quantitative, visual, and verbal form. The explanations are created to be interpretable even for nonexperts. On top of the diagnostic mode for explanations, the system offers an educational mode.

> ExAID offers two complementary operation modes that are meant for different use cases. A diagnostic mode provides functionality meant to support dermatologists during the clinical examination of patients' skin lesions. For research and education purposes, ExAID offers an educational mode including a collection of tools for holistic analysis of the deep model's behaviour as well as the collected case data.
>
> (Lucieri et al. 2022, 4)

This approach aligns with recent calls within the biomedical research field about educating end users to understand AI-powered technologies better (Fritzsche et al. 2023) and tackling the literacy component recognized as essential to enhance transparency.[12]

Investigations on social aspects of AI adoption in biomedical research, particularly in biobanking, have highlighted the potential of co-design "as a strategy to improve fairness, accountability, and transparency of algorithmic systems" (Donia and Shaw 2021, 1). In the ExAID case, the co-designing strategy involved a wide range of 35 experts (philosophers, ethicists, policymakers, social scientists, medical doctors, legal and data protection specialists, computer scientists, and machine learning (ML) engineers, two of them representing the patients) (Zicari et al. 2021b). Involving a wide range of stakeholders is consistent with the suggestions earlier discussed[13] about the potential impact of engaging external stakeholders in the co-design process on reducing asymmetries in access to information.

By discussing different socio-technical scenarios, the multidisciplinary team identified that multiple usages for the same AI technology were possible in this early design phase. The team considered "the pros and cons and evaluate[d] which of the various possible usages is the primary use case" (Zicari et al. 2021b, 5). Recognizing the various paths technology can take at the early stages of the innovation process also means accepting that "technology trajectory"[14] can vary following the influence of determined "relevant social groups"[15] and societal conditions and requirements pending over the innovation process. A co-design strategy also recognizes the "negotiability" of technological development, which refers to "the scope for particular groups and forces to shape technologies to their ends and the possibility of different kinds of ('technological' and 'social' outcome)" (Williams and Edge 1996, 867). Embracing this approach contributes to advancing a nondeterministic path to AI development and reinforces the democratic and open governance of technology.

As another critical outcome, the workshops conducted during the co-design shed light on the potential scopes of the tool and the priorities and concerns of different stakeholders. As a result, the team detected tensions between the perspectives of various stakeholders, for example, dermatologists and evidence-based medicine experts, as Table 8.1[16] shows.

Table 8.1 Dermatologists and Evidence-Based Medicine Expert Feedback

Argument 4	*Counterargument*
Viewpoint: The Dermatologist. Early detection of malignant melanoma is critical, as the risk of metastasis with worse prognosis increases the longer melanoma remains untreated.	*Viewpoint:* The Evidence-Based Medicine. There are no reliable biomarkers that can predict the prognosis of melanoma before excision. There are patients who survive their localized melanoma without therapy. Therefore, the early diagnosis does not necessarily mean a better prognosis; on the contrary, there is a risk of poor patient care due to overdiagnosis.

The various perspectives that a co-design strategy provides can help AI designers and data curators embed multiple perspectives into a particular AI-driven tool's "technical code".[17] In the case of the ExAID, their heterogeneity allowed the team to confront "differences and tensions" (Zikari et al. 2021b). The developers acknowledge how the Z-inspection® process allowed them to become aware of relevant tensions for the further development and potential deployment of the prototype. "This includes the definition of the target metric, which usually differs a lot from a machine learning perspective, compared to clinical and evidence-based medicine point of view (Number needed to treat, vs. Mortality Rate, vs. Survival Rate)".[18]

The case illustrates a possible path to follow while using a co-design approach, which can result in enhanced transparency of the innovation and adoption processes. As in the case of HistoMapr™, the path followed by the ExAID team emphasizes interpretability by addressing understandable communication through intelligible language and engagement practices. This approach shows the potential impact of interdisciplinary stakeholder teams on building trust in AI systems.

The ExAID project and these proposals remind us of the relevance of embracing a broader approach to the issues of AI transparency and explainability because it allows the development of engagement strategies for different actors. Knowing the tensions that might emerge among different groups relevant to technology development can also serve to design clear explanations to educate and anticipate questions emerging from different scenarios.

3 Discussion

As we have seen, adopting AI-driven tools makes the communication of processes and results in biomedical research more complex due to the difficulties in understanding how the systems took their decisions and explained them in interpretable ways. As discussed, the lack of explainability impacts the overall model transparency and can negatively impact trust in biobanking (Kargl et al. 2022). Enhancing transparency and making the systems more interpretable[19] are critical to achieving an "ecosystem of trust"[20] for AI adoption in biobanking.

Scholars[21] have been raising awareness of the need to bring social sciences learnings into the AI context of what transparency and explainability mean and to develop broader frameworks of understanding. Proposals developed within the social, health, and cognitive sciences discussed help to think about alternatives that can make the functionality and outcomes of the model interpretable to various stakeholders involved in the technology development, including end users.

This session discusses critical learning drawn from the analysis of the cases and the review of the literature, which can contribute to understanding AI transparency and explainability in their communicative dimension. These ideas can only cover some of the challenges that adopting AI-powered tools entails for biomedical research from a communicative perspective. The intention is to contribute to advancing the conversation on AI transparency from the social sciences perspective, particularly from a communication standpoint. Although the application of these ideas does not constitute independent silos in practice, on the contrary, they are deeply intertwined; for a better understanding, they are stated as follows: (1) understanding transparency as a feature of the innovation and adoption processes; (2) embracing co-design strategies to enhance transparency, and (3) incorporating built-in and external solutions to provide interpretable explanations.

By sharing these ideas, this chapter proposes that communicative efforts toward transparency and explainability can complement efforts within the computational sciences to make the systems explainable (when possible) and the transition to interpretable models for high-stake decisions.

3.1 Understanding Transparency as a Feature of the Innovation and Adoption Processes

Understanding transparency as a communication issue requires recognizing it as a feature of AI innovation and adoption processes. To deepen this perspective, it is helpful to embrace transparency as "openness". This proposal builds on the discussion about the multifaceted character of transparency (Oxelheim and Forssbæck 2014) and is particularly associated with the concept of open government (Larsson and Heintz 2020). In this discussion, transparency comes with "a political framing of empowerment under the notion of fostering democratic processes" (Larsson and Heintz 2020, 6). Acknowledging this perspective is essential for at least two reasons: a model cannot build transparency by itself because AI "transparency" is a socio-technical construct of artifacts and humans (Kemper and Kolkman 2019), and more transparent processes in AI adoption have the potential to impact the democratic governance of biobanks.

From a communication perspective, embracing this approach means paying attention to the asymmetries in accessing information about the data and model functionality and unpacking different transparency stances. This perspective entails a differentiated approach to data, algorithmic, and performance transparency (Matheny et al. 2019), and considers different scenarios and functions (e.g., providing transparency by design or intentional explanation of the model deployment) (Loi et al. 2021). As these proposals remind us of the granular character of transparency, this issue helps us design strategies for the various actors and aspects involved in developing AI-driven technologies. Data is one of its most critical components, "thus, the transparency of the available data such that others can critically evaluate the training process is paramount to ensuring confidence in the system" (Kozlakidis 2020, 200). Nevertheless, it is critical to flag that disclosing which data was used to train the model to determine stakeholders enhances data transparency but does not necessarily make the whole process transparent. For this reason, understanding transparency as a feature of the entire processes of innovation and adoption requires seeing its granular character, paying attention to every part that might influence the whole, and considering its holistic nature.

Another critical aspect of this discussion is considering the differentiated needs for transparency. This idea ponders the differentiated communication needs of the various actors involved in AI development. Learning from the use of AI-driven tools for PRSs, these actors might include: "patients, relatives of patients, various public(s), healthcare professionals, medical/nursing students, researchers, technicians, ethics committees, clinical ethics teams, business partners and all the other stakeholders involved in research and development as well as the translation, and implementation" (Fritzsche et al. 2023, 7) of these technologies.

Recognizing the needs and roles of different actors involved in the socio-technical assemblage of AI development allows us to return to the notion of transparency associated with reducing information asymmetries (Oxelheim and Forssbæck 2014) mentioned earlier. Embracing the communicative dimension of transparency also means acknowledging the necessity of providing meaningful information tailored according to the needs of every actor and the content to discuss. However, as the case of exAID illustrates, the democratic potential of transparency will not be fulfilled; at least those processes are participative.

3.2 Embracing Co-Design Strategies to Enhance Transparency

The potential of participatory and engagement mechanisms with end users has been recognized as a valuable strategy for co-designing AI-powered technologies for health and biomedical research (Donia and Shaw 2021). The example of ExAID illustrates how conversations that support two-way learning[22] as a mechanism of engagement can effectively contribute to bringing new perspectives that can be out of the radar of AI engineers. In this case, the workshops, including end users (dermatologists and patient representatives), helped redefine the initial purpose and scope of the technology. Every actor brought a unique perspective to the discussion after sharing relevant information on the tool. Another learning from this case is the challenging nature of the interdisciplinary conversation. Guided by the Z-Inspection® (Zicari et al. 2021a) framework, researchers questioned the initial aim of the tool according to different standpoints of expertise. The process details the complex character of co-design mechanisms due to the challenging nature of interdisciplinary research. "Successful interdisciplinary research is challenging and requires participants to articulate in detail exactly how concepts – including the most basic – are theoretically and practically understood" (Zicari et al. 2021b, 5). Then, adopting a co-design strategy requires considering a translational approach among disciplines, expertise, and ways of life to make the dialogue relevant to all the actors. The development of AI-powered tools for biobanking could, for instance, include nonexpert public representatives providing inputs on:

i. The data collection and curation processes (e.g., identifying potential biases or gaps that researchers may overlook and selection criteria).
ii. The development of bot-driven consent forms.[23]
iii. The type of predictions a multitask model[24] can make by using data representative of a particular population or disease.
iv. The development of AI-powered tools for researching potential participants (see chapter "Artificial Intelligence and Communication With Research Participants in the Process of Biobanking Human Biological Material" by Jarosław Sak and Jakub Pawlikowski).

The investigation on AI-driven PRSs "strongly recommends adopting interactive/participatory engagement practices, especially between clinicians and patients" (Fritzsche et al. 2023, 6). The scope of co-design strategies might include a wide range of actors and purposes. The case of ExAID shows that "stakeholder engagement should not be restricted to end-user involvement but needs to encourage and support the inclusion of additional stakeholder groups" (Robertson et al. 2019, 592). This perspective can contribute to embedding those multiple perspectives into the "technical code" of the AI-driven tools for biobanking and strengthening transparency in governance.

In the case of ExAID, the discussion encouraged the team "to involve patients and representative patient groups more prominently in the next stages of the development, and to more openly communicate/document important limitations and risks of the application" (Adriano Lucieri, email message to author, 4 April 2023). The main impact of the co-design process was "in the direction of data privacy" (Adriano Lucieri, email message to author, 4 April 2023), as the team currently works on privacy-preserving machine learning methods for medical image analysis while investigating the impact of privacy attacks on their system. This experience shows how a participatory approach also entails recognizing the active role citizen-patients (Fritzsche et al. 2023) might have in developing technology for digital health. This approach acknowledges that challenges in science and technology are more than technological or scientific dilemmas[25] and have the potential to expose their social and political nature. This means approaching issues of transparency and explainability of AI from a situated point of view, considering the broader socio-economic structure where the tools will be used/deployed. While adopting engagement mechanisms as co-design strategies, the imperative of avoiding functionalists and deficit communication models is also a key learning brought about by years of study of engagement/involvement in biobanking (Luna et al. 2020).

As research has shown, most "biobank studies report public involvement in a 'functionalist' way, with the purpose of improving recruitment processes and participation rates" (Luna et al. 2020, 760). Most of the initiatives were intended to enhance consent forms and recruitment processes, and those studies sought confirmation or support for a determined project using the public's opinion. Still, some empirical evidence in the past suggests that the public might be willing to participate in biobanking differently (Terry et al. 2012). This fact may be particularly interesting for initiatives aimed at developing AI technologies for biobanking because of the unexpected outcomes a co-design strategy might provide, particularly at the early stages of the innovation process, as the ExAID project shows. While the functionalist character of engagement mainly looks for quantifiable outcomes, a socio-constructivist approach will consider the contextual situation and participants' backgrounds and facilitate a two-way learning process, which can benefit the achievement of more transparent AI innovation and adoption.

From the literature on forms of engagement with biobanks, the most common are focus groups and deliberative exercises, ad-hoc consultation and support, community-based participatory research, public representatives in the biobank governance structure, and surveys (Luna et al. 2020). Those forms and previous learnings in biobank experiences can also contribute to developing and implementing co-design strategies for AI technologies. Findings from the mechanisms of engagement in the British Columbia (BC) Generations Project[26] can guide helpful principles adapted to the co-design process. Among those findings emerged the need for "having mechanisms that allow for continuing communication between management and the donor collective as a whole" (O'Doherty et al. 2011, 372), which encourages the adoption of two-way and continuous communication among stakeholders.

Whether the co-design strategy is intended to involve the public, other relevant actors, or both relevant to the technology development, acknowledging the benefits of the participatory approach and its socio-technical character is critical to contributing to the transparency of AI innovation and adoption processes in its governance. ExAID shows that the outcome of a co-design process can span from restating the tool's aim, foreseeing potential socio-technical scenarios, including their challenges, obtaining inputs on potential harms or benefits from various standpoints, and identifying the best forms of communicating about the model outcomes and performance, as seen in the next section.

Involving diverse actors concerned with the development of the technology, including end users, might contribute to achieving a more transparent process and obtaining insights on how to create intelligible explanations based on context and situation. This idea learns from a long-standing tradition in communication studies and studies of technology and science (STS), where constructivist approaches have called attention to the interpretative character of technology[27] and, consequently, to the conditions pending over the social shaping of technology.

3.3 Incorporating Systems Built-in and External Solutions to Provide Interpretable Explanations

Experiences in the broad scenario of health sciences and the case of HistoMapr™ show the impact that intelligible explanations, tailored to the needs and expertise of concerned actors, have on the overall transparency of AI tools. Ideally, a co-design process can contribute to determining the main concerns of the actors involved and then tackle those questions by using built-in or external solutions to provide interpretable explanations. This means that achieving those pieces of explanation (whatever the form they take) can be planned as a desired outcome of the co-design process. For example, feedback from groups of patients can provide clues on how to communicate potential risks or concerns about a particular technology to a broader audience. The experiences discussed in the chapter allow one to conclude that explanations of AI-driven tools can take the form of add-ons or external solutions in the form of communicative or educational products.

Exploring the influence of communication methods for an app that predicts skin cancer, researchers found that "a simple one-page visual or sentence that explains the criteria or process used to make medical decisions increased acceptance of an algorithm-based skin cancer diagnostic tool" (Cadario et al. 2021, 1640). The conception and implementation of those explanations can be based on cognitive and social science learning regarding the character contrastive and social interactions of "everyday explanations" (Miller 2019) discussed earlier.

However, it is essential to make a critical distinction. Providing interpretable explanations about the performance and outcomes of an AI-driven tool should not be taken as a synonym with public relations about a particular artifact. The same investigation on skin cancer tools found that in their marketing efforts, tech companies tend to emphasize terms like accuracy and convenience while providing few details about how algorithms work (Cadario et al. 2021). Unlike that style, the current proposal favors transparent communication focused on promoting user engagement through those explanations as part of broader efforts to promote transparency, as explained previously.

Following a transparent communication style instead of a marketing approach is consistent with the recommendations discussed, which are anchored in social and cognitive sciences. It is linked to the potential of critical communication to enhance transparency.

Those communication artifacts could then consider the characteristics of particular contexts, how users in those scenarios interpret the main concepts shared, and provide a clear explanation of the relationship between risk and benefit, adopting an AI literacy approach. HistoMapr™ provides an excellent example of explanations grounded in the needs of the final users, in that case, pathologists. In coherence with recognizing transparency as a socio-technical construct of people and artifacts, this proposal learns from a long tradition in Science Communication and Cultural Studies that highlights the role of context, situation, and individual in the reception of any communicative act.

The taxonomy of types of explanations on AI proposed by the Alan Institute and ICO guidelines for "Explaining explanations made with AI" helps think of the different stances of the innovation and adoption processes about the functions of those explanations.

Rationale explanation: The reasons that led to a decision delivered in an accessible and nontechnical way.

Responsibility explanation: Who is involved in the development, management, and implementation of an AI system, and who to contact for a human review of a decision?

Data explanation: What data has been used in a particular decision and how?

Fairness explanation: Steps taken across the design and implementation of an AI system to ensure that the decisions it supports are generally unbiased and fair and whether or not an individual has been treated equitably.

Safety and performance explanation: Steps taken across the design and implementation of an AI system to maximize the accuracy, reliability, security, and robustness of its decisions and behaviors.

Impact explanation: Steps taken across the design and implementation of an AI system to consider and monitor the impacts that the use of an AI system and its decisions has or may have on an individual and on wider society (ICO and The Alan Turing Institute 2020).

Table 8.2 Box of Questions for Communicating AI4H

Type of Explanation	*Example of Questions*
Rationale Explanation	i. Why do we need this tool in the first place? ii. What changes in the diagnosis of this disease with this technology? iii. What is the main purpose of this tool?
Responsibility Explanation	iv. Who is responsible for the impact of this tool? Who is behind it?
Data Explanation	v. What type of data was used to train the model, and from which sources? vi. How vulnerable is the AI system against (training) data leakage? What safety measures are in place to avoid (training) data leakage? How secure is the general handling of data transfer during inference?
Fairness Explanation	vii. Did this set include patients/families like me? viii. What are the risks of bias? ix. How balanced was the training population?
Safety and Performance Explanation	x. Does it provide multiple outcomes? xi. Can we explain how it makes its decisions? xii. Is there any risk to my health in its use? xiii. What is the scope of the tool? What is it not able to do/process?
Impact Explanation	xiv. What are the benefits? xv. What do I need to know to use it? Is any specific training required?

Using various methods and communicative forms to apply these taxonomy of explanations aligns with transparency's granular and situated character. Considering the distinct functions and scenarios, those explanations can lead to enhanced interpretability. Following these stances of explanations, and as a part of a broader framework for AI transparency from a communication standpoint, this chapter proposes a Box of questions for communicating AI4H created by the author that can be adapted to the setting of biomedical research institutions (see Table 8.2[28]).

Answering these questions or some of them, according to the public, through visual or textual, interactive or noninteractive products could help organizations embrace complementary explainability methods from a communication perspective. Using a language anchored to the users' needs would also promote AI literacy.

4 Conclusion

In this chapter, I have discussed how social approaches to transparency and explainability can contribute to confronting some of the challenges that adopting AI-driven technologies entails for biomedical research. The literature review focused on the social dimensions of transparency, explainability, and interpretability principles, allows one to conclude that there is a necessity to bring the epistemic understanding of social and cognitive sciences to the solutions for explainability in the context of AI development. The analysis of the AI-powered tools "HistoMapr™ (rebranded as PredxBio Platform)" and "ExAID" development, interpreted against the backdrop of learnings from public engagement and transparency in biomedical research, helps to confirm the convenience of embracing communicative-participatory approaches to enhance transparency in the adoption of AI-driven technologies. Those approaches can take the form of co-design strategies, interpretable explanations tailored to the needs of end users, and interpretable salience methods or communicative products. As transparency has been linked to trust, more openness can potentially lead to enhanced public trust in AI-driven technologies. Dialogic communicational approaches still have strong potential to contribute to the democratic governance of AI in biomedical research.

Notes

1 For further references on the intersections between accountability, oversight, and transparency in biobanking, see Deschênes and Sallée (2005); Gille et al. (2020); and Laurie et al. (2018).

2 In an email dated 14 October 2023, Dr. Akif Burak Tosun, Director of Product @ PredxBio, explained that HistoMapr was rebranded to PredxBio Platform and now it also presents explainable results for multiplexed images of tissue samples, not only limited to transmitted light pathology images. This chapter will keep its original analysis referred to HistorMap.

3 For further discussion on the transparency and explainability as neighboring concepts, see Christian, "The Alignment Problem" and Goisauf and Cano, "Ethics of AI in Radiology".

4 Debates within computer sciences call for a shift from explainability to interpretability, warning about the efforts of going against the nature of deep neural networks to try to make those black boxes explainable. Instead, Cynthia Rudin calls for policymakers "not to accept black box models without significant attempts at interpretable (rather than explainable) models" (2019, 211). This approach comes with several arguments to explain why explainable black boxes should be avoided in high-stakes decisions and potentially replaced with interpretable models. As Brian Christian notes, confronted with the dilemma of large, unintelligible versus explainable models, it is valid to ask "whether we ought to be using large or complex models in the first place" (2021, 90).

5 Shapley is one of the methods used in xAI, it shows the contribution of each feature to the prediction.

6 LIME is a popular method used in xAI that provides explanations for individual predictions made by a machine learning model.

7 Miller (2019) offers a comprehensive analysis of how transparency has been understood in cognitive and social sciences.

8 Miller (2019) argues that "everyday explanations address 'why particular facts (events, properties, decisions, etc.) occurred', rather than general scientific relationships".

9 See in the Literature Review section, the discussion on studies conducted by Miller (2019), "Explanation in Artificial Intelligence", and Mittelstadt et al. (2019), "Explaining Explanations in AI".

10 The philosophical roots of this discussion are in Marcuse's (1982, 41) notion of formal rationality while analyzing modern technology and society relationships.

11 Further discussion on the formal rationality influencing AI development can be found at Elliott (2019).

12 For example, the WHO (2021) "Ethics and Governance of Artificial Intelligence for Health" recognizes information-sharing and literacy as critical to transparency in AI.

13 For example, see proposals by Fritzsche et al. (2023) and Zicari et al. (2021a).

14 Within Studies of Science and Technology, the socio-constructivist approach known as social shaping of technology (SST) proposes the notion of technology trajectory which

states that there is not one single, predetermined trajectory for innovation, but multiple potential paths influenced by various factors. See MacKenzie and Wajcman (1999).

15 The notion of relevant social groups is core to the theoretical approach known as the Social Construction of Technology. It tackles the groups that influenced the development of technology artifacts. See Pinch et al. (1984).

16 Zicari et al. (2021b) share different groups' arguments and counterarguments. The different views in same matters demonstrate the relevance of involving diverse stakeholders in order to embed various perspectives in the design.

17 Technical code is a notion proposed by Feenberg (2002). It suggests that the technology design embedded interests and characteristics of the economic and political arrangements.

18 Adriano Lucieri, email message to author, 4 April 2023.

19 The WHO (2021) acknowledges the links between transparency on the information about the systems and interpretability.

20 The EU AI Act emphasizes strongly on the need of creating an ecosystem of trust for AI.

21 See studies discussed in the literature review: Fritzsche et al. (2023); Goisauf and Abadía (2022); Larsson and Heintz (2020); Miller (2019); Mittelstadt et al. (2019).

22 Participatory processes should embrace a mutual learning approach (Luna et al. 2020).

23 The development of these consent forms can take into consideration the approach proposed by O'Doherty et al. (2011) and the principles for solid biobanking governance.

24 Christian (2021) discusses the potential and challenges of multitask AI models for health.

25 See Bucchi and Trench (2008) for further discussions on the social character of scientific controversies.

26 The project is led by researchers at British Columbia University Cancer and has hundreds of data variables for research into the causes of cancer and chronic disease. See O'Doherty et al. (2011) for further details.

27 See more on the interpretable character of technology in Bijker (2015).

28 The methodology and workflow to apply the Box of Questions for Communicating AI4H are currently being developed for future publication; the idea is to offer guidance on how to adopt it to enhance transparency of AI systems toward end users. The author acknowledges the contribution of Dr. Adriano Lucieri with questions vi, ix, and xiii.

Reference List

Bijker, Wiebe E. 2015. "Social Construction of Technology." In *International Encyclopedia of the Social & Behavioral Sciences (Second Edition)*, edited by James D. Wright. Oxford: Elsevier. https://doi.org/10.1016/B978-0-08-097086-8.85038-2.

Bucchi, Massimiano, and Brian Trench. 2008. "Genetics and Genomics: The Politics and Ethics of Metaphorical Framing." In *Handbook of Public Communication of Science and Technology*, edited by Massimiano Bucchi and Brian Trench. Routledge.

Cadario, Romain, Chiara Longoni, and Carey K. Morewedge. 2021. "Understanding, Explaining, and Utilizing Medical Artificial Intelligence." *Nature Human Behaviour* 5 (12): 1636–42. https://doi.org/10.1038/s41562-021-01146-0.

Christian, Brian. 2021. *The Alignment Problem: How Can Machines Learn Human Values?* London: Atlantic Books.

Deschênes, M., and C. Sallée. 2005. "Accountability in Population Biobanking: Comparative Approaches." *The Journal of Law, Medicine & Ethics* 33: 40–53.

Dive, Lisa, Christine Critchley, Margaret Otlowski, et al. 2020. "Public Trust and Global Biobank Networks." *BMC Medical Ethics* 21 (1): 73–81. https://doi.org/10.1186/s12910-020-00515-0.

Donia, Joseph, and James A. Shaw. 2021. "Co-Design and Ethical Artificial Intelligence for Health: An Agenda for Critical Research and Practice." *Big Data & Society* 8 (2). https://doi.org/10.1177/20539517211065248.

Elliott, Anthony. 2019. *The Culture of AI: Everyday Life and the Digital Revolution*. London: Routledge. https://doi.org/10.4324/9781315387185.

European Commission. 2019. Policy and Investment Recommendations for Trustworthy Artificial Intelligence, Shaping Europe's Digital Future (AI HLEG). https://digital-strategy.ec.europa.eu/en/library/policy-and-investment-recommendations-trustworthy-artificial-intelligence (08.04.2025).

European Commission. 2020. White Paper of the European Commission of 19 February 2020 on Artificial Intelligence – A European Approach to Excellence and Trust COM(2020) 65 final.

European Commission. 2021. *Proposal for a Regulation of The European Parliament and of the Council Laying Down Harmonised Rules on Artificial Intelligence (Artificial Intelligence Act) and Amending Certain Union Legislative Acts*. https://eur-lex.europa.eu/legal-content/EN/TXT/?uri=CELEX:52021PC0206 (08.04.2024).

European Commission. Directorate-General for Communications Networks, Content and Technology. 2019. Ethics Guidelines for Trustworthy AI. Publications Office.

Feenberg, Andrew. 2002. *Transforming Technology: A Critical Theory Revisited*. Oxford University Press. http://ebookcentral.proquest.com/lib/ucalgary-ebooks/detail.action?docID=430829.

Fritzsche, Marie-Christine, Kaya Akyüz, Mónica Cano Abadía, Stuart McLennan, Pekka Marttinen, Michaela Th. Mayrhofer, and Alena M. Buyx. 2023. "Ethical Layering in AI-Driven Polygenic Risk Scores – New Complexities, New Challenges." *Frontiers in Genetics* 14. https://doi.org/10.3389/fgene.2023.1098439.

Garvey, Colin, and Chandler Maskal. 2020. "Sentiment Analysis of the News Media on Artificial Intelligence Does Not Support Claims of Negative Bias Against Artificial Intelligence." *OMICS: A Journal of Integrative Biology* 24 (5): 286–99. https://doi.org/10.1089/omi.2019.0078.

Gille, Felix, Effy Vayena, and Alessandro Blasimme. 2020. "Future-Proofing Biobanks' Governance." *European Journal of Human Genetics* 28 (8): 989–96. https://doi.org/10.1038/s41431-020-0646-4.

Goisauf, Melanie, and Mónica Cano Abadía. 2022. "Ethics of AI in Radiology: A Review of Ethical and Societal Implications." *Frontiers in Big Data* 5. https://doi.org/10.3389/fdata.2022.850383.

Gyevnar, Balint, Nick Ferguson, and Burkhard Schafer. 2023. "Bridging the Transparency Gap: What Can Explainable AI Learn from the AI Act?" In *ECAI 2023*, 964–71. IOS Press. https://doi.org/10.3233/FAIA230367.

ICO, and The Alan Turing Institute. 2020. "Explaining Decisions Made with AI." https://ico.org.uk/for-organisations/uk-gdpr-guidance-and-resources/artificial-intelligence/explaining-decisions-made-with-artificial-intelligence/ (08.04.2025).

Jobin, Anna, Marcello Ienca, and Effy Vayena. 2019. "The Global Landscape of AI Ethics Guidelines." *Nature Machine Intelligence* 1 (9): 389–99. https://doi.org/10.1038/s42256-019-0088-2.

Kargl, Michaela, Markus Plass, and Heimo Müller. 2022. "A Literature Review on Ethics for AI in Biomedical Research and Biobanking." *Yearbook of Medical Informatics* 31 (1): 152–60. https://doi.org/10.1055/s-0042-1742516.

Kemper, Jakko, and Daan Kolkman. 2019. "Transparent to Whom? No Algorithmic Accountability Without a Critical Audience." *Information, Communication & Society* 22 (14): 2081–96. https://doi.org/10.1080/1369118X.2018.1477967.

Kozlakidis, Zisis. 2020. "Biobanks and Biobank-Based Artificial Intelligence (AI) Implementation Through an International Lens." In *Artificial Intelligence and Machine Learning for Digital Pathology: State-of-the-Art and Future Challenges*, edited by Andreas Holzinger, Randy Goebel, Michael Mengel, and Heimo Müller, 195–203. Cham: Springer International Publishing. https://doi.org/10.1007/978-3-030-50402-1.

Larsson, Stefan, and Fredrik Heintz. 2020. "Transparency in Artificial Intelligence." *Internet Policy Review* 9 (2). https://doi.org/10.14763/2020.2.1469. https://policyreview.info/concepts/transparency-artificial-intelligence.

Laurie, G. T., E. S. Dove, A. Ganguli-Mitra, I. Fletcher, C. McMillan, N. Sethi, and A. Sorbie. 2018. "Charting Regulatory Stewardship in Health Research: Making the Invisible Visible." *Cambridge Quarterly of Healthcare Ethics* 27 (2): 333–47. https://doi.org/10.1017/s0963180117000664.

Lipton, Zachary C. 2017. "The Mythos of Model Interpretability." *arXiv*. https://doi.org/10.48550/arXiv.1606.03490.

Loi, Michele, Andrea Ferrario, and Eleonora Viganò. 2021. "Transparency as Design Publicity: Explaining and Justifying Inscrutable Algorithms." *Ethics and Information Technology* 23 (3). https://doi.org/10.1007/s10676-020-09564-w.

Lucieri, Adriano, Muhammad Naseer Bajwa, Stephan Alexander Braun, Muhammad Imran Malik, Andreas Dengel, and Sheraz Ahmed. 2022. "ExAID: A Multimodal Explanation Framework for Computer-Aided Diagnosis of Skin Lesions." *Computer Methods and Programs in Biomedicine* 215 (2): 106620. https://doi.org/10.1016/j.cmpb.2022.106620.

Lucieri, Adriano, Muhammad Naseer Bajwa, Andreas Dengel, and Sheraz Ahmed. 2020. "Explaining AI-Based Decision Support Systems Using Concept Localization Maps." In *Neural Information Processing*, edited by Haiqin Yang, Kitsuchart Pasupa, Andrew Chi-Sing Leung, James T. Kwok, Jonathan H. Chan and Irwin King, 185–93. Communications in Computer and Information Science. Cham: Springer International Publishing, https://doi.org/10.1007/978-3-030-63820-7_21.

Luna Puerta, L., W. Kendall, B. Davies, S. Day, and H. Ward. 2020. "The Reported Impact of Public Involvement in Biobanks: A Scoping Review." *Health Expect* 23 (4): 759–88. https://doi.org/10.1111/hex.13067.

MacKenzie, Donald, and Judy Wajcman. 1999. *The Social Shaping of Technology*. 2nd ed. Buckingham: Open University Press.

Marcuse, Herbert. 1982. "Some Implications of Modern Technology." In *The Essential Frankfurt School Reader*, edited by Andrew Arato and Gebhardt Eike. New York: Continuum. http://archive.org/details/essenti_xxx_1982_00_9819.

Matheny, Michael, Sonoo Thadaney Israni, Mahnoor Ahmed, and Danielle Whicher, eds. 2019. *Artificial Intelligence in Health Care. The Hope, the Hype, the Promise, the Peril*. Washington, DC: National Academy of Medicine. https://nam.edu/wp-content/uploads/2021/07/4.3-AI-in-Health-Care-title-authors-summary.pdf.

Miller, Tim. 2019. "Explanation in Artificial Intelligence: Insights from the Social Sciences." *Artificial Intelligence* 267: 1–38. https://doi.org/10.1016/j.artint.2018.07.007.

Mittelstadt, Brent, Chris Russell, and Sandra Wachter. 2019. "Explaining Explanations in AI." *Proceedings of the Conference on Fairness, Accountability, and Transparency, Atlanta, GA, USA, Association for Computing Machinery*: 279–88. https://doi.org/10.1145/3287560.3287574.

O'Doherty, K. C., M. M. Burgess, K. Edwards, et al. 2011. "From Consent to Institutions: Designing Adaptive Governance for Genomic Biobanks." *Social Science & Medicine* 73 (3): 367–74. https://doi.org/10.1016/j.socscimed.2011.05.046.

Oxelheim, Lars, and Jens Forssbæck. 2014. "The Multifaceted Concept of Transparency." In *The Oxford Handbook of Economic and Institutional Transparency*, edited by Lars Oxelheim and Jens Forssbæck. Oxford University Press.

Pinch, Trevor, and Wiebe Bijker. 1984. "The Social Construction of Facts and Artefacts: Or How the Sociology of Science and the Sociology of Technology Might Benefit Each Other." *Social Studies of Science* 14 (3): 399–441.

PredxBio. 2023. "Technology." https://predxbio.com/technology.

Robertson, L. J., R. Abbas, G. Alici, A. Munoz, and K. Michael. 2019. "Engineering-Based Design Methodology for Embedding Ethics in Autonomous Robots." *Proceedings of the IEEE* 107 (3): 582–99.

Rudin, Cynthia. 2019. "Stop Explaining Black Box Machine Learning Models for High Stakes Decisions and Use Interpretable Models Instead." *Nature Machine Intelligence* 1 (5): 206–15. https://doi.org/10.1038/s42256-019-0048-x.

Slunecka, J. L., M. D. van der Zee, J. J. Beck, B. N. Johnson, C. T. Finnicum, R. Pool, et al. (2021). "Implementation and Implications for Polygenic Risk Scores in Healthcare." *Human Genomics* 15 (1): 46. https://doi.org/10.1186/s40246-021-00339-y.

Terry, Sharon F., Kurt D. Christensen, Susan Metosky, et al. 2012. "Community Engagement about Genetic Variation Research." *Population Health Management* 15 (2): 78–89. https://doi.org/10.1089/pop.2011.0013.

Tosun, Akif Burak, Filippo Pullara, Michael J. Becich, D. Lansing Taylor, S. Chakra Chennubhotla, and Jeffrey L. Fine. 2020. "HistoMaprTM: An Explainable AI (XAI) Platform for Computational Pathology Solutions." In *Artificial Intelligence and Machine Learning for Digital Pathology: State-of-the-Art and Future Challenges*, edited by Andreas Holzinger, Randy Goebel, Michael Mengel and Heimo Müller. Cham: Springer International Publishing.

Webber, Jutta, and Bianca Prietl. 2022. "AI in the Age of Technoscience: On the Rise of AI-Data Driven and Its Episteme-Ontological Foundations." In *The Routledge Social Science Handbook of AI*, edited by A. Elliot. Routledge.

Williams, Robin, and David Edge. 1996. "The Social Shaping of Technology." *Research Policy* 25: 865–99.

World Health Organization (WHO). 2021. "Ethics and Governance of Artificial Intelligence for Health. WHO Guidance." Geneva: World Health Organization. https://iris.who.int/handle/10665/349093.

Zicari, Roberto V., Sheraz Ahmed, Julia Amann, et al. 2021b. "Co-Design of a Trustworthy AI System in Healthcare: Deep Learning Based Skin Lesion Classifier." *Frontiers in Human Dynamics* 3: 2. www.frontiersin.org/articles/10.3389/fhumd.2021.688152.

Zicari, Roberto V., John Brodersen, James Brusseau, et al. 2021a. "Z-Inspection®: A Process to Assess Trustworthy AI." *IEEE Transactions on Technology and Society* 2 (2): 83–97. https://doi.org/10.1109/TTS.2021.3066209.

Part 4

Embracing the Potentiality

9 Social Acceptance of Artificial Intelligence in Biobanking

Elodie Caboux, Patrick Courtney, Io Hong Cheong, and Zisis Kozlakidis

1 Introduction

The technological developments of recent decades have heralded computer-driven approaches in the development of laboratory research, clinical practice, biobanking, and the ongoing automation of their operations (Burnham et al. 2013). The implementation of computer-based approaches in healthcare and biomedical research, able to handle and analyze clinical tests and records is not a new one. It was discussed since the late 1950s concurrently with some of the very first implementations within a healthcare environment (Fish 1958; Casscells et al. 1978). However, those discussions were carried through with little conviction (other than for administrative purposes) until the last three decades and the understanding that computational approaches need to be both strengthened and integrated within healthcare and biomedical research. Flagship international projects, such as the Human Genome Project, acted as the catalyst in transforming our understanding and expectations of computer-driven approaches in healthcare (Roses 2000), and were followed by considerable investment to allow for such approaches to emerge.

At the same time, the rapid pace of technological development has meant that the information obtained from biospecimens (such as the commonly collected blood and tissue samples, and urine) can be of much greater detail and variety than ever before (Nachmias and Linn 1987; Thurow et al. 2004; Murdoch et al. 2013). For example, in the past decade alone, clinical microbiology has experienced revolutionary advances in terms of culture-independent molecular detection assays, laboratory automation, information systems linkage, and point-of-care testing, to name but a few (Vandenberg et al. 2018). A similar trajectory has been recorded for the developments in clinical imaging (Holzinger et al. 2019; Larson et al. 2020). Thus, the information in the different biospecimens can now be readily and routinely extracted through new technological approaches, leading to vast amounts of digital data being captured and stored in machine-readable forms. However, this increasing complexity of clinically relevant information must be captured in meaningful and actionable ways to support biomedical and clinical research, and treatment(s). The applications of machine learning (ML) techniques in medicine allow the more efficient handling of these vast data volumes (often collectively

DOI: 10.4324/9781032634487-14

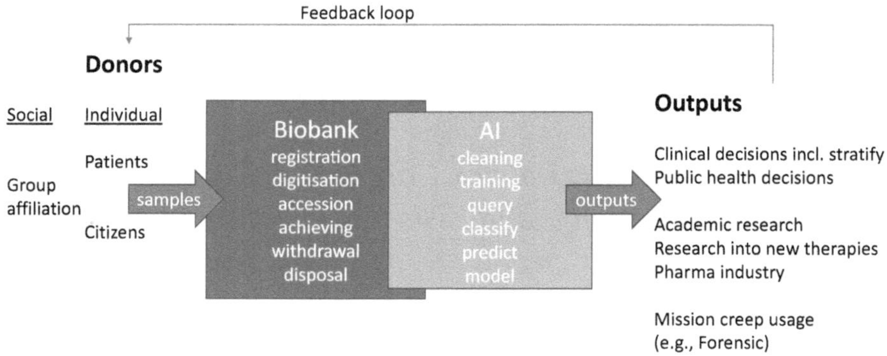

Figure 9.1 The simplified, schematic relationship between the wider social environment, individuals, biobanks, digitization and outputs, as considered within the current chapter.

termed as "big data"), to interpret the data and inform decision-support systems with the so-called artificial intelligence (AI). AI is a vague but evocative expression to denote, in this context, the capabilities of machines (i.e., algorithms) to classify or stratify clinical cases or predict related conditions with high reliability (sensitivity and specificity) – in some cases, even more accurately than human experts, and potentially reducing bias and human errors (Cabitza et al. 2020). Figure 9.1 represents the interrelationship between AI and biobanking, as considered by the authors in this chapter.

Taking the earlier into account, biobanks are being relied upon to standardize tissue collection at scale for improved science quality and to drive research, as they systematically collect large amounts of samples, supporting their subsequent "datafication" and clinical research. In biobanking specifically, datafication contains two distinct aspects, firstly, the deep collection of demographic and healthcare data, and secondly, the collection of molecular, imaging, and omics data, following the analyses of these collected samples (Thorogood 2020). In doing so, tissue-based science is also becoming an informatics-driven challenge (Kozlakidis 2016). The implementation of AI in biobanking has been presented previously in some detail (Holzinger 2015; Kozlakidis 2020). A number of high-profile translational research initiatives globally are based on the availability of big infrastructural units (100,000 Genomes Project and Our Future Health in the UK; the Precision Medicine Initiative in the USA) that would support the eventual technological transfer of high-throughput analytical approaches, such as AI, from the research into the (consolidated) biobanking facility and subsequently (or in some cases in parallel) to the clinical laboratory (Kozlakidis 2020).

However, the ethical aspects of this interaction between biobanks and AI remain still under intense consideration (Jongsma and Bredenoord 2020; Sorell et al. 2022; Rosemann and Zhang 2022; Gille et al. 2020). The current work aims to review the efforts made to consider the ethics – in particular social acceptance – of AI in the field of biobanking. The available evidence, relating to the ethics of AI-driven applications in biobanking, is reviewed from an international perspective, and is not restricted to a particular application or geographical location. For clarification purposes, some key terms are defined in Section 2 followed by Section 3 discussing the social acceptance of technology (such as AI) in biomedical research. Section 4 will highlight how these aspects cross over and translate into biobanking, while Section 5 will present the challenges that the clinical, pharmaceutical, and academic research communities are facing, while providing an outline to future outlooks.

2 Glossary and Key Terms

Social acceptance is defined as the degree to which a phenomenon (e.g., AI implementation in healthcare and/or biobanking) is accepted by relevant social actors, based on the degree to which the phenomenon is (dis-)liked by these actors (Von Wartburg and Liew 1999). Inherently this definition of social acceptance is a confluence between two dimensions: social acceptance (an empirical fact) and the ethical judgment of a technology (Van de Poel 2016).

3 Methodology

A narrative review approach has been chosen to investigate the topic of AI and social acceptance in biobanking, as the aim was to provide a broad perspective and explore the general debates/developments on the topic, rather than implementing a systematic review approach, as the latter focuses on unique and specific queries, utilizing explicit methodology (Ferrari 2015). The search terms "biobank" OR "biobanking" AND "artificial intelligence" AND "social acceptance" were used in the databases PubMed and Web of Science since inception, to identify the published literature pool. However, no manuscripts were identified – this was not surprising as this is a novel subject of investigation. Thus, an alternate search was performed using the terms "social acceptance" AND "biobank" OR "biobanking", resulting in 283 identified manuscripts. After filtering out conference proceedings, poster presentations, and non-English publications, the remaining abstracts were individually investigated for an inclusion of big data handling and artificial intelligence, resulting in 23 manuscripts identified. The remaining abstracts were indexed by theme to the three aspects of sociopolitical, community, and market acceptance, by the two authors (EC and IHC) and validated independently by the third one (ZK). Primary references were used as a source to derive further related references to ensure, as much as possible, thematic saturation.

4 Social Acceptance and Artificial Intelligence in Medical Research

The successful implementation of new technologies, in any field, requires their ethical acceptance; social acceptance is a constituent part of this ethical umbrella view. Historically, the implementation of new technologies in biomedical research was considered a relatively simple issue that i) had to happen to maintain efficiencies in the face of increased demands (Gelijns and Halm 1991; Laal 2012) and ii) was considered a relatively simple issue that could be addressed by applying the appropriate communication strategies to raise awareness or showcasing technical prowess (Shaw 1985; see also chapters "Artificial Intelligence and Communication With Research Participants in the Process of Biobanking Human Biological Material" by Jarosław Sak and Jakub Pawlikowski and "Unique by Design – the Need for Individualized Models of Trust in Biobanking AI" by Anna Clareborn). As such, the widely perceived inevitability of progress meant that any challenges with wider acceptance could be overcome by formal education (Keane and Topol 2021) and community engagement. However, such an approach lacks the fundamental understanding of the different steps (and actions) that describe the process of acceptance; starting with raising overall awareness of an innovation to the desired target audience, such that the innovation is tested/used, and following this with appropriate education to enable the necessary culture change within the biomedical environment, so that the use of the innovation becomes supported and eventually entrenched (Hastall et al. 2017) or in the case of a negative outcome, rejected. Furthermore, the wider social acceptance is primarily negatively defined (following the definitions from other technological fields) as nontechnical factors/aspects (Wüstenhagen et al. 2007).

It is important at this point to also highlight the current simplistic view regarding the sample donor/research participant and its status in AI implementation in biobanking. In most publications, this status is described as or assumed to be a passive one, that is, sample donors are passive receptors of knowledge, often a considerable time after their sample and data donation to a biobank. The same approach is evident also in the implications of potential failure of the tools, such as AI, supporting research that utilizes biobanked data. The impacts are considered on an individual level, for example, lack of explainability of AI methodology, and not in their wider social context. Yet, it is well established from other technological innovations that acceptance is both a passive and active process, with different components affecting each aspect (Hastall et al. 2017; Dopson et al. 2008). Specifically, the passive acceptance indicates the mere reception of the technology or technology output, while the active acceptance means being willing to accept the technology, interact with it, and actively recommend it to others (so that it becomes adopted widely). There is of course negative advocacy too, where groups of individuals may be using a technology (e.g., 5G in telecommunications), yet be positioned against it. However, the overall experience from the COVID-19 pandemic has demonstrated that patients, patient groups, and citizens at large are actively interacting with technology, through many different digital applications (Abusanad 2021; Pardes et al. 2022; Jazieh et al. 2020). Thus, it would be inconsistent if this passive view were to remain for AI implementation in biobanking. Instead, the previous examples demonstrate the need to prepare for an active engagement with sample donors/patients, including steps beyond a basic communication, such as raising awareness of the technological potential and educating them on the value and process of biobanking.

Without underestimating the need or impact of education regarding the social acceptance of AI in biomedical research, it is yet appropriate to take a more holistic view when we consider social acceptance. Studies on social acceptance carried out in other rapidly evolving technological fields (e.g., renewable energies) distinguish three dimensions of social acceptance, all with their own characteristics and momentum (Van de Poel 2016; Forsythe 2001; Wolsink 2012). As transcribed in biomedical research, they would be: a) sociopolitical acceptance, with a focus on the frameworks that set the foundation for the other two dimensions; b) community acceptance, which concerns both the professional and lay communities (citizens, and patients [treated, recovered, chronic], and biobank donors); and c) market acceptance, which is about the willingness among different actors to invest/pay for these new technologies, such as AI. Each one of those aspects will be reviewed briefly next, to compose a holistic view of social acceptance and AI in biomedical research.

4.1 Sociopolitical Acceptance

Innovation is the successful application of new ideas, transferred into products, processes, or organizations. The integration of new technologies such as AI in biomedical research is not a matter of simply applying science; there exist factors in the social environment that co-determine it: social, economic, cultural, and different technological factors. This is why simply making a new technology affordable does not automatically guarantee its acceptance as the other factors also need to be satisfied (Yildirim et al. 2021). Specifically in biomedical research, almost all AI innovation in healthcare is focusing on new capabilities, for example, researchers have created systems that predict if a patient will die (Song et al. 2018); will be discharged (Zhang et al. 2020); and predict the onset of conditions like sepsis (Nemati et al. 2018); or the patient outcome post-discharge from an Intensive care unit (ICU) (Last et al. 2016). However, the assumptions made by many AI implementations often indicate a lack of understanding of clinical practice workflows (Topol 2019) or consider a very narrow targeting of "textbook" cases (for which normally the best training datasets exist), and for which the clinicians would be able to identify and act upon anyway (Yang et al. 2015). The challenge exists in that the impact of an erroneous decision might be inconvenient, ineffective, or in extreme cases even fatal for a patient, and as such subject to liability repercussions.

Yet, one of the main clinical daily challenges is the existence of continuous high volume of alerts and prompts (e.g., for the automated reporting of biochemical analyses of blood samples) that distract clinicians from their work, and that provide a negative orientation to spending more time with patients (Rajkomar et al. 2019; Singh et al. 2013). As such, the strong business case for the AI implementation is missing, added to the *de facto* absence of empathy by AI implementation (Fulmer 2019) – even if from a research perspective published implementation cases remain interesting. Additionally, the medical environment typically operates within a complex network of legacy informatics systems, and thus any further additions increase the interoperability, complexity (Yildirim et al. 2021; Reisman 2017; Glaser 2020), and associated costs (including purchase costs, staff time for trainers/training, and testing/onboarding time). Finally, there is little research published, detailing the challenges for clinical decision-making and thus identifying the time and place of highest impact for an AI implementation (Glaser 2020). As such professional groups, such as clinicians, biomedical researchers, and associations of medical disciplines, remain skeptical in terms of overall acceptance (Cai et al. 2019).

4.2 Community Acceptance

Community acceptance is often seen as a bottleneck, and a large part of the existing literature focuses on specific populations or patient groups (Lee et al. 2018; Capstick et al. 2009; Yang Meier et al. 2020; Vu and Lim 2021). Most published works show an undisputable support for medical and pharmaceutical computer-driven innovations, whose benefits are acknowledged by a high majority of respondents and interviewed people in qualitative research studies. For most communities, healthcare remains a main priority, and as such any technical solution providing an improvement would be welcomed in principle. Furthermore, the tight regulation of the medical profession provides a layer of trust, in that any new implementation would be within existing, tried, and tested frameworks. At the same time, this tight regulation has been highlighted as a potential context-dependent barrier to innovation adoption and diffusion (Mittra and Tait 2012).

However, community acceptance issues arise clearly when AI implementation, for example, utilizing genomic information, entails the possibility for discriminating people (e.g., for treatment options and life insurance) because people are not equal with respect to their vulnerability to diseases and other stresses (Hoffman 2021; Cirillo et al. 2020). This AI-driven potential for bias or "algorithmic discrimination" forms a core element in community acceptance of AI implementation in biomedical research, as it directly affects the aspects of fairness and privacy (Hoffman and Podgurski 2019). This bias may be further amplified by the many types of data input that due to low quality and/or lack of completeness and representation, can produce outputs that can entrench such a potential discrimination. For example, the training of AI skin cancer detection algorithms, that took place almost exclusively with images of white skin, eventually underperforming for other skin tones (Noor 2020; see also chapter "Fairness of an AI System in the Case of a Biobank of Images and Imaging Biomarkers" by Valentina Colcelli).

Additionally, a major element in community acceptance is that of explainability (Holzinger et al. 2019), and its constituent part of interpretability (i.e., how understandable the outputs of AI implementation are) for the diverse stakeholders. Thus, clinicians being able to interpret the outcomes of AI implementation are more likely to successfully explain them (Markus et al. 2021). AI implementations are necessary to reenact the machine decision-making process, to reproduce and to comprehend both the learning and knowledge extraction process. This is important, because for the eventual clinical decision-support, it is necessary to understand the causality of learned representations (Gershman et al. 2015). Furthermore, AI explainability enhances the trust of professionals and community alike in AI systems (Meacham et al. 2019; Knapič et al. 2021; Confalonieri et al. 2021), however, one has to consider that different explanations/metaphors/models may need to be utilized to communicate effectively to different types of stakeholders. At this point, it has to be pointed out that there are different types of explainability and that not all AI can be explained. For example, there are effective Deep Learning implementations that cannot be explained, and other implementations that are fully transparent but less accurate (Holzinger et al. 2019; Bologna and Hayashi 2017), leading to feelings of loss of control by clinicians for implementations that cannot be explained (Reis et al. 2020). As such, a tension will remain inherent to community acceptance, which needs to be identified and communicated as clearly as possible.

The earlier aspects described the community acceptance from the perspective of the end user. However, in biobanking, a considerable portion of the community is constituted by the donors of biological material themselves, be they patients or healthy volunteers. Several studies have focused on community acceptance in biobanking, including the intensive use of data, and the emerging consensus is that there is a general support for research that will address healthcare challenges. However, donors also express their desire for active decision-making (e.g., being recontacted for additional consent) and the need for a greater awareness of the potential uses of data in research. Having said that, there is not a clear consensus on the types of research uses that would be acceptable and can result in consistent mission creep as described in Information Technology. This probably demonstrates the varied understanding of the subject, thus highlighting the need for further awareness and education regarding the implementation of AI in biomedical research and biobanking in particular (Thiel et al. 2014; Spruill et al. 2014; Sanderson et al. 2017).

4.3 Market Acceptance

The market acceptance relates directly to the willingness of different actors to invest in AI implementation in biomedical and clinical research. Indeed, the motivation of different actors to invest would vary widely. For example, public actors (e.g., public fund holders, granters, and foundations) might be more willing to invest in the case where there is not an immediate profit, but long-term savings at a population level, while for private actors a more immediate return on investment might be warranted. There exist investment models applied to AI in other fields, for example, winner takes all, which are likely to continue influencing the decision-making process of private actors when it comes to AI in biomedical and clinical research (Naudé and Dimitri 2020).

From an institutional standpoint, investment in AI is generally considered a strategic decision, as it can be both considerable and impactful. Having said that, recent systematic work has shown that the existing financial impact assessments on AI implementation show methodological deficits. For example, often alternatives to achieve similar impact are not evaluated in parallel, hence evading a comprehensive comparison (Wolff et al. 2020). From a national/international standpoint, the economic investment in AI implementation in biomedical research relies on

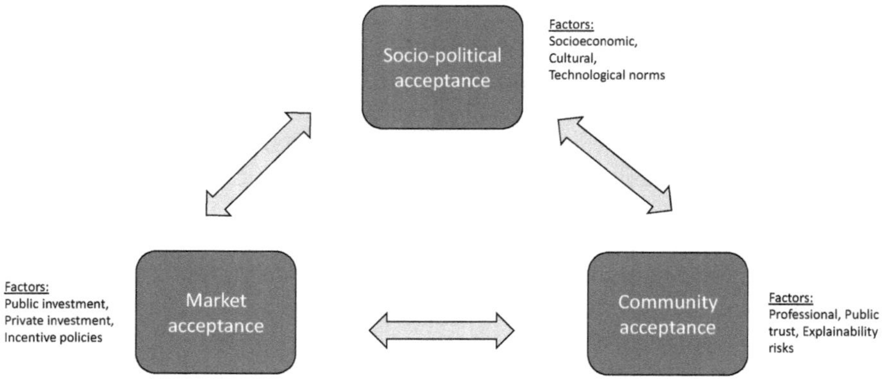

Figure 9.2 The three main factors of social acceptance as relating to the introduction of new technologies.

policy instruments that are usually designed (or perceived as designed) to create incentives for positive investment decisions (Roberts et al. 2021; Owoyemi et al. 2020). In general, as the translation of biomedical research output into clinical care might be a lengthy process, the need for such policy instruments and incentives cannot be overlooked. As the economic investment in healthcare is markedly different between High-income and Low-and-Middle-Income Countries (LMICs), AI implementation might also become an additional piece of evidence (e.g., in addition to physical equipment) reflecting this infrastructural gap (WHO 2021b). Figure 9.2 provides a schematic summary of these three social acceptance factors.

5 Social Acceptance and Artificial Intelligence in Biobanking

The discussions on AI implementation in the fields of translational medicine, biomedical research, and biobanking are not new and have been presented previously in some detail (Holzinger et al. 2015). Collecting clinical specimens for future research use and combining those with extensive clinical, diagnostic, geographic, and demographic data in biobanks is essential for the development of improved diagnostic tools and the eventual improvement of patient care. The NIH Human Microbiome Project, which served as a catalyst for human microbiome research, highlights the added value of large consolidated biobanking (NIH Human Microbiome Portfolio Analysis Team 2019). Therefore, the implementation of AI-driven techniques promises to lessen the level of expertise required of bench technologists and biobanking staff alike to perform repetitive tasks rapidly, accurately, and consistently. Indeed, it is expected that AI will ultimately facilitate high-throughput biobanking in a cost- and time-effective manner, responding in practice to the increasing volumes of biological samples, the multidisciplinary character, and the complexity of biobanking processes, and will facilitate downstream abilities to query databases, improve standardization (due to the scalability), and thus potential interoperability between biobanks.

While there has been consistent research on the social acceptance of biobanks in general (NIH Human Microbiome Portfolio Analysis Team 2019; Robertson 2003; Gottweis and Zatloukal 2007; Rothstein 2005; Goisauf et al. 2019), within specific geographic locations, cultural, and ethnic contexts (Igbe et al. 2012; Khatib et al. 2021; Amin et al. 2018; de Vries et al. 2017), and patient/population groups (Schneider et al. 2016; Van Assche et al. 2013; Raivola et al. 2019), the social acceptance of biobanking in relation to AI has not been investigated adequately yet. The indications that exist currently from a sociopolitical acceptance perspective are again focusing mostly on the technological aspects rather than a more holistic view (O'Sullivan et al. 2017; Vayena et al. 2018). As such the acceptance of AI in biobanking is considered as part of the overall convergence of technologies within biomedical research, and not on a separate footing. Having said that, biobanks themselves are facing a dual bottleneck: specifically, to align with the expanding technological requirements and clinical research needs and become "AI-ready", there is increased pressure to harmonize both their analytical laboratory and digital techniques (data collection, curation, and storage). Both aspects directly affect any AI-driven initiatives and can significantly impact biobank utilization rates and long-term biobank sustainability (Kozlakidis 2016). This intense biobanking focus is firmly on the technical aspects, thus it is not surprising that this is reflected on the social acceptance discourse as well.

With regard to community acceptance, there is still insufficient granularity in considering the consistency of the communities themselves. For example, professional communities that interact with patients are likely to have different opinions from those that are laboratory-based and have no contact. Similarly, the wider public is too broad a term including different types of patients (current, recovered, and chronically managed), as well as citizens, however, for current reasons of simplicity, these communities are often presented as large aggregates. Considering this diversity, there have been distinct voices by the professional and patient communities requesting further ethical exploration of biobanking aspects. For example, in the past decade, a new type of biobanks has emerged: "living" biobanks or organoid biobanks, based on the older expertise of tissue-type culture collections that created the HeLa cells (Lucey et al. 2009) and other pluripotent, human-derived cell lines. In short, human tissues are transformed into biotechnological artifacts that can grow, expand, and be stored limitlessly, with some organoids being effectively immortal, and some potentially acquiring considerable scientific, clinical, and commercial value (Boers et al. 2016; Calandrini et al. 2020; Sachs et al. 2018). With the implementation of new technologies, such as AI, and the open-ended nature of biobanking-facilitated research, conflicting interests and ethical challenges for the stakeholders (e.g., donors, patients, researchers, clinicians, industry, and the wider public) will arise. Moreover, organoids may acquire distinct types of moral value, particularly if they resemble human tissues for which research use is ethically controversial and under different legal jurisdictions, such as brain organoids, gonadal organoids, and self-organizing embryo models (e.g., gastruloids) that mimic early human embryonic development (Rivron et al. 2018). As a result, the professional community themselves have raised the need for a proactive ethical consideration of the field (Bredenoord et al. 2017; Boers and Bredenoord 2018; Jongsma and Bredenoord 2020). This indicates that while the need has been identified, a level of community acceptance has not yet been reached, and there currently exist different regional viewpoints, for example, from the USA, EU, and other global regions.

Lastly, with regard to market acceptance, the benefit of implementing AI within biobanking has been framed as a cost-saving operation (as opposed to an added-value proposition) (Bowton et al. 2014). For example, implementing AI in biobanked datasets improves drug candidate molecules design and as such reduces the overall costs of drug development (Chen et al. 2021), and perhaps offers the potential to lower the cost for orphan drugs (Schreidah et al. 2023). While these arguments might be correct, and perhaps necessary to drive the initial investment by institutions, they represent only one aspect. For example, added-value propositions can be achieved through AI implementation in biobanks embedded within clinical facilities. The structured nature of the biobank data (e.g., including images of tissues) is amenable to AI processing and can be used to answer biomedical and/or clinical research questions, as well as the added value of inferring trends in public health and related costs from the same data (Alberich-Bayarri et al. 2019). Thus, the degree of market acceptance remains limited.

6 Challenges and Outlook

The issue of social acceptance of AI in biobanking will become even more prominent in the coming decade, as more biobanks modernize and enhance their digital capacities, as free-standing infrastructures are, more likely, embedded in an increasingly digital healthcare system. There is also the possibility of federated structures, where the algorithm may be sent to the infrastructure to process the data. At the same time, the character of acceptance in the three dimensions will probably change. The current challenge of institutional adoption and initial investment constraints (i.e., the early adoption phase of the new technologies) will gradually shift toward social acceptance, where the emphasis is placed on the connectivity and the ethical risks emerging thereof. In other words, the argument will move from introducing a new type of analytical tool toward sufficient capacity and resourced toward a sustainable integration into routine operations. While this is a natural progression that has been observed in other fields (e.g., renewable energy, telecommunications, and manufacturing), the impact of such a natural shift tends to be underestimated within national healthcare policy planning, where decisions on innovation adoption are often delegated to a decentralized structure (e.g., the Integrated Care Boards in the UK National Health Service). Hence, there is commonly a delay in creating uniform, nationwide incentives that will drive toward a positive market acceptance. At this point, it is important to note that not being able to reach social acceptance of AI implementation in biobanking is likely to have severe and detrimental effects on the potential use of data for biomedical research as a whole, affecting primarily the institutions that have invested heavily in its development, being unable to materialize a return on that investment. In the case of AI and biobanks, most of the funding comes from the public purse.

A relevant example was seen in the case of genetically modified organisms (and their potential implementation in agriculture), where social acceptance was highly diverse, leading to their banning in many parts of the world where it was lowest, and subsequent limited growth in the field even when new technologies were becoming available (Cho 2016; Varzakas et al. 2007).

Specifically, for the sociopolitical acceptance of AI implementation in biobanking, the challenges of "explainability" and "interpretability" of these technologies remain (Varzakas et al. 2007). This is particularly true when self-learning applications are taken into consideration. Self-learning ability is defined as a certain degree of self-programming, after evaluating big data deriving from a biobank, which includes the relevance of existing data to specific clinical contexts. Compared to conventional computer applications, these are processes of data appraisal beyond direct human control. Although the algorithms developed for such systems are evidence-based, their automatic outcomes escape by default from the area of knowledge of both the biomedical researchers as well as the system developers (Patrzyk et al. 2017; Institute of Electrical and Electronics Engineers 2017) – though no AI can exceed the limits of its own program. As such the aspects of transparency and trust can be directly affected (see chapter "Tackling AI Transparency Concerns in Biomedical Research: Bringing a Communication-Participatory Approach to the Conversation" by Leslie Salgado Arzuaga). There is also another ethical question here: whether the use of these systems can meet the informed consent requirements (as an expression of personal autonomy), raising further the question on transparency regarding their functional characteristics (Garattini et al. 2019). Furthermore, the AI implementation in biobanking predisposes the need for collecting and sharing data at scale, thus highlighting the challenge of scalability on ethical, legal, and governance aspects of biobanking. As such, the notions of informed consent and data protection would need to be extended to facilitate the AI-driven solutions in biobanking (Beauchamp and Childress 2001; Char et al. 2018; Paul et al. 2017).

In terms of community acceptance, reviews of ethical studies have shown that there is a general positive attitude of the public toward biobanking (after the term and operations have been explained by a trained individual) – however, the perceptions of potential harm varied widely (D'Abramo et al. 2015). Among the factors influencing the degree of acceptance were the potential collaboration of biobank research with the pharmaceutical industry, and insurance industries, and the participants' trade-off between privacy and perceived utility. Indeed, the protection of individual privacy and communal rights stands in apparent contradiction to the gathering of samples and data in biobanks. On the other hand, the role of the individual has also been emphasized, for example, in the form of regulations protecting individual rights (i.e., privacy), with the result that scientific practices, such as population registries, are challenged because of new laws emphasizing the use of informed consent for scientific practices, which are beneficial for local communities (Casali 2014). Furthermore, a recent study in Ghana, Uganda, and Zambia under the auspices of the Human Heredity and Health in Africa initiative (H3Africa) suggested that while biobank participants are willing to give consent for their samples and associated data to be used for future research purposes, they equally expect to receive feedback and have access to the information about the progress of the research and about the kinds of research being undertaken on their samples and data (Tindana et al. 2020). The aspect of reciprocity is not a new observation in biobanking (Haimes and Whong-Barr 2004; Gaskell et al. 2013), though not expressed equally as strongly globally (Antonova and Eritsyan 2022). As such, these expectations relating to AI implementation in biobanking should be anticipated and discussed as part of an ongoing communication process to achieve community acceptance in the long term. Furthermore, it is likely that as AI is implemented at different rates and within different contexts, such discussions on benefit sharing might diverge between high-income and low-and-middle-income countries (as in the latter, the relative prioritization of healthcare needs will differ) (WHO 2021a).

Regarding the professional community, the risk of bias remains in any AI implementation, including in the field of biobanking. Bias in outcomes arises from bias inherent in the training data (e.g., well-annotated, standardized data) and/or design choice of the algorithm and has been observed in previous AI implementations (Fenech et al. 2018). It is already well-recognized that both healthcare and evidence-based medicine are biased against disadvantaged groups, as these groups are underrepresented in the evidence base (Rogers 2004) (including in biobanks) (Keyes and Westreich 2019). Having said that, solutions can be tested and implemented. For example, a team from MIT's Computer Science and Artificial Intelligence Laboratory (CSAIL) and Massachusetts General Hospital (MGH) presented a deep-learning mammogram algorithm specifically designed to perform equally well for American women of African and Caucasian descent (Yala et al. 2019) by carefully curating the input training data. Thus, this aspect needs to become part of the discussion of professional acceptance and replace the often-mentioned argument of "more data is needed", when it is not available, not relevant, and not demonstrated.

Lastly, in terms of market acceptance, AI implementation in biobanking is likely to continue to be strongly dependent on the national and/or regional context within which biobanking facilities need to operate. As observed in other fields, the speed with which this happens relates to the stakeholders involved in the process, including (at a minimum) the institutions, the biobanks, the staff, as well as the algorithm developers (companies, charities, and foundations). Thus, technical aspects such as modularity and scalability (e.g., integration of multiple data sources) become important design considerations for manufacturers. Additionally, the incorporation of AI into biobank operations will also be dependent on the availability of skilled staff within geographic regions.

The earlier observations directly impact the transparency and accountability of ethical aspects (Banks 2018, see also chapter "Powerful AI: Between Accountability and Potentiality" by Michaela Th. Mayrhofer). For example, if AI is implemented remotely through a virtual connection to a biobank that is geographically remote, where does the accountability lie if an autonomous AI system makes a mistake? While several actors are involved in this example (e.g., the publisher of the software solution; the developer; the buyer – such as the laboratory/institution; and the user(s) – such as the doctor or the biobank), there might not be a sole attribution of accountability to any of the earlier, but rather different relative proportions/contributions to the creation of the mistake. Finally, it is unknown as to what extent the regulatory desire for control of samples and data (Chen and Song 2018) might influence future AI market acceptance.

The current narrative review explores a novel subject area and as such it might contain some inherent limitations. Specifically, most manuscripts identified focused on biomedical research and social acceptance of AI implementation, with biobanking being an adjunct or associative operation. Thus, there is a bias in the viewpoint of clinical implementation and market acceptance, coming primarily from clinical staff servicing an additional research role, and not research staff that have a dedicated biobanking role. In contrast, most of the community acceptance manuscripts are derived from qualitative research driven by biobanks. Thus, there exist divergent views that have not been investigated further as this is out of the scope of the current chapter. Furthermore, the model of social acceptance has been transcribed from other rapidly developing technological areas, where it has been previously successfully applied – yet even in that field exceptions on social acceptance exist, for example, the 5G controversy. The literature search has been performed in English, and as such may have missed viewpoints from other languages where there is a recorded skepticism of AI implementation, for example, in German. Having said that, it might still require further fine-tuning that is specific to the biobanking and biomedical research field.

7 Conclusion

While AI is implemented gradually into biomedical, public health, and clinical research and in biobanking, it is not yet part of routine operations. However, the path for a more regular future use has been already set. As the available databases in biobanks grow, processing this information will become challenging. AI solutions are already on the market and there is increasing evidence of the efficacy of AI medical software in clinical decision support. Therefore, the ethical aspects of AI implementation in biobanking merit further investigation, at a minimum, to lay the ground for an anticipated routine use.

The current narrative review has provided an overview of the social acceptance context of AI in biomedical research, before delving deeper into the field of biobanking. A model was used for social acceptance, transcribed from other actively growing technological fields, and consisting of three aspects: sociopolitical, community, and market acceptance. The AI implementation in biobanking was considered based on this model, reviewing the status, as well as the many remaining open challenges for the future.

In conclusion, it can be said that the social acceptance of AI alone contains several gaps and opportunities on all three dimensions of social acceptance. The opportunities for AI implementation in biobanking are of significant impact (e.g., creating clinical decision-support systems, and providing research answers to complex clinical questions) and thus there seems to be a positive correlation with social acceptance overall. In contrast, gaps include the lack of examples for context-driven implementation (and benefits thereof) within biobanking that would enhance professional acceptance. The challenges of explainability and interpretability of AI implementation constitute challenges affecting the sociopolitical acceptance, while the market acceptance is highly sensitive to regulatory and governmental incentives, thus creating a fragmented outlook for the field. When combined with further technological advances and the collections of biological samples and data at scale, these gaps are further enhanced by additional aspects impacting autonomy, privacy, explainability, accountability, market, and community acceptance. Thus, the need arises for further work on the subject in a proactive manner to facilitate the projected benefits (or in the case of failure, adapt/alter course) of AI implementation for biomedical research, biobanking, and the common good.

Disclaimer

Where authors are identified as personnel of the International Agency for Research on Cancer/WHO, the authors alone are responsible for the views expressed in this chapter and they do not necessarily represent the decisions, policies, or views of the International Agency for Research on Cancer/WHO.

Acknowledgments

The authors would like to acknowledge the BBMRI-ERIC ELSI Team for fruitful and insightful discussions over several occasions that have given rise to some of the thoughts included in this work.

Reference List

Abusanad, Atlal. 2021. "'Najia' Story: A WhatsApp Support Group for Patients with Breast Cancer." *Innovations in Digital Health, Diagnostics, and Biomarkers* 1 (1): 16–18.

Alberich-Bayarri, Angel, Emanuele Neri, and Luis Martí-Bonmatí. 2019. "Imaging Biomarkers and Imaging Biobanks." In *Artificial Intelligence in Medical Imaging*, edited by Erik R. Ranschaert, Sergey Morozov and Paul R. Agra, 119–26. Cham: Springer.

Amin, Latifah, Hasrizul Hashim, Zurina Mahadi, and Khaidzir Ismail. 2018. "Determinants of the Willingness to Participate in Biobanking Among Malaysian Stakeholders in the Klang Valley." *BMC Medical Research Methodology* 18: 1–14.

Antonova, Natalia, and Ksenia Eritsyan. 2022. "It Is Not a Big Deal: A Qualitative Study of Clinical Biobank Donation Experience and Motives." *BMC Medical Ethics* 23 (1): 1–11.

Banks, Jim. 2018. "The Human Touch: Practical and Ethical Implications of Putting AI and Robotics to Work for Patients." *IEEE Pulse* 9 (3): 15–18.

Beauchamp, Tom L., and James F. Childress. 2001 *Principles of Biomedical Ethics*. Oxford University Press.

Boers, Sarah N., and Annelien L. Bredenoord. 2018. "Consent for Governance in the Ethical Use of Organoids." *Nature Cell Biology* 20 (6): 642–5.

Boers, Sarah N., Johannes J. M. van Delden, Hans Clevers, and Annelien L. Bredenoord. 2016. "Organoid Biobanking: Identifying the Ethics: Organoids Revive Old and Raise

New Ethical Challenges for Basic Research and Therapeutic Use." *EMBO Reports* 17 (7): 938–41.

Bologna, Guido, and Yoichi Hayashi. 2017. "Characterization of Symbolic Rules Embedded in Deep DIMLP Networks: A Challenge to Transparency of Deep Learning." *Journal of Artificial Intelligence and Soft Computing Research* 7 (4): 265–86.

Bowton, Erica, Julie R. Field, Sunny Wang, et al. 2014. "Biobanks and Electronic Medical Records: Enabling Cost-Effective Research." *Science Translational Medicine* 6 (234): 234cm3.

Bredenoord, Annelien L., Hans Clevers, and Juergen A. Knoblich. 2017. "Human Tissues in a Dish: The Research and Ethical Implications of Organoid Technology." *Science* 355 (6322): eaaf9414.

Burnham, Carey-Ann D., W. Michael Dunne Jr., Gilbert Greub, Susan M. Novak, and Robin Patel. 2013. "Automation in the Clinical Microbiology Laboratory." *Clinical Chemistry* 59 (12): 1696–702. https://doi.org/10.1373/clinchem.2012.201038.

Cabitza, Federico, Andrea Campagner, and Clara Balsano. 2020. "Bridging the 'Last Mile' Gap Between AI Implementation and Operation: 'Data Awareness' That Matters." *Annals of Translational Medicine* 8 (7).

Cai, Carrie J., Samantha Winter, David Steiner, Lauren Wilcox, and Michael Terry. 2019. "'Hello AI': Uncovering the Onboarding Needs of Medical Practitioners for Human-AI Collaborative Decision-Making." *Proceedings of the ACM on Human-Computer Interaction* 3: 1–24.

Calandrini, Camilla, Frans Schutgens, Rurika Oka, Thanasis Margaritis, et al. 2020. "An Organoid Biobank for Childhood Kidney Cancers That Captures Disease and Tissue Heterogeneity." *Nature Communications* 11 (1): 1310.

Capstick, Stuart, Pauline Norris, Faafetai Sopoaga, and Wale Tobata. 2009. "Relationships Between Health and Culture in Polynesia–A Review." *Social Science & Medicine* 68 (7): 1341–8.

Casali, P. G. 2014. "Risks of the New EU Data Protection Regulation: An ESMO Position Paper Endorsed by the European Oncology Community." *Annals of Oncology* 25 (8): 1458–61.

Casscells, Ward, Arno Schoenberger, and Thomas B. Graboys. 1978. "Interpretation by Physicians of Clinical Laboratory Results." *New England Journal of Medicine* 299 (18): 999–1001.

Char, Danton S., Nigam H. Shah, and David Magnus. 2018. "Implementing Machine Learning in Health Care – Addressing Ethical Challenges." *The New England Journal of Medicine* 378 (11): 981.

Chen, Yongxi, and Lingqiao Song. 2018. "China: Concurring Regulation of Cross-Border Genomic Data Sharing for Statist Control and Individual Protection." *Human Genetics* 137 (8): 605–15.

Chen, Zhaoyi, Xiong Liu, William Hogan, Elizabeth Shenkman, and Jiang Bian. 2021. "Applications of Artificial Intelligence in Drug Development Using Real-World Data." *Drug Discovery Today* 26 (5): 1256–64.

Cho, Hang-Min. 2016. "Study of Trend of Reporting on GMO as a Convergence Technology: Focused on Analyzing Articles in Major Domestic Daily Newspapers from 1994 to 2015." *Journal of Digital Convergence* 14 (12): 267–81.

Cirillo, Davide, Silvina Catuara-Solarz, Czuee Morey, et al. 2020. "Sex and Gender Differences and Biases in Artificial Intelligence for Biomedicine and Healthcare." *NPJ Digital Medicine* 3 (1): 81.

Confalonieri, Roberto, Ludovik Coba, Benedikt Wagner, and Tarek R. Besold. 2021. "A Historical Perspective of Explainable Artificial Intelligence." *Wiley Interdisciplinary Reviews: Data Mining and Knowledge Discovery* 11 (1): e1391.

D'Abramo, Flavio, Jan Schildmann, and Jochen Vollmann. 2015. "Research Participants' Perceptions and Views on Consent for Biobank Research: A Review of Empirical Data and Ethical Analysis." *BMC Medical Ethics* 16 (1): 1–11.

de Vries, Jantina, Syntia Nchangwi Munung, Alice Matimba, et al. 2017. "Regulation of Genomic and Biobanking Research in Africa: A Content Analysis of Ethics Guidelines, Policies and Procedures from 22 African Countries." *BMC Medical Ethics* 18: 1–9.

Dopson, Sue, Louise Fitzgerald, and Ewan Ferlie. 2008. "Understanding Change and Innovation in Healthcare Settings: Reconceptualizing the Active Role of Context." *Journal of Change Management* 8 (3–4): 213–31.

Fenech, Matthew, Nika Strukelj, and Olly Buston. 2018. "Ethical, Social, and Political Challenges of Artificial Intelligence in Health." United Kingdom. https://wellcome.org/sites/default/files/ai-in-health-ethical-social-political-challenges.pdf (12 12 2024).

Ferrari, Rossella. 2015. "Writing Narrative Style Literature Reviews." *Medical Writing* 24 (4): 230–5.

Fish, B. R. 1958. "Applications of an Analog Computer to Analys of Distribution and Excretion Data." *Health Physics* 1 (3): 276–81.

Forsythe, Diana. 2001. *Studying Those Who Study Us: An Anthropologist in the World of Artificial Intelligence*. Stanford University Press.

Fulmer, Russell. 2019. "Artificial Intelligence and Counseling: Four Levels of Implementation." *Theory & Psychology* 29 (6): 807–19.

Garattini, Chiara, Jade Raffle, Dewi N. Aisyah, Felicity Sartain, and Zisis Kozlakidis. 2019. "Big Data Analytics, Infectious Diseases and Associated Ethical Impacts." *Philosophy & Technology* 32: 69–85.

Gaskell, G., H. Gottweis, J. Starkbaum, et al. 2013. "Publics and Biobanks: Pan-European Diversity and the Challenge of Responsible Innovation." *European Journal of Human Genetics: EJHG* 21 (1): 14–20. https://doi.org/10.1038/ejhg.2012.104.

Gelijns, Annetine C., and Ethan A. Halm. 1991 "An Introduction to the Changing Economics of Technological Innovation in Medicine." In *The Changing Economics of Medical Technology,* edited by Annetine C. Geljins and Ethan A. Halm. Washington, DC: National Academies Press (US).

Gershman, Samuel J., Eric J. Horvitz, and Joshua B. Tenenbaum. 2015. "Computational Rationality: A Converging Paradigm for Intelligence in Brains, Minds, and Machines." *Science* 349 (6245): 273–8.

Gille, Felix, Effy Vayena, and Alessandro Blasimme. 2020. "Future-Proofing Biobanks' Governance." *European Journal of Human Genetics* 28 (8): 989–96.

Glaser, John. 2020. "It's Time for a New Kind of Electronic Health Record." *Harvard Business Review*. https://hbr.org/2020/06/its-time-for-a-new-kind-of-electronic-health-record?shtml.

Goisauf, Melanie, Gillian Martin, Heidi Beate Bentzen, et al. 2019. "Data in Question: A Survey of European Biobank Professionals on Ethical, Legal and Societal Challenges of Biobank Research." *PLoS ONE* 14 (9): e0221496.

Gottweis, Herbert, and Kurt Zatloukal. 2007. "Biobank Governance: Trends and Perspectives." *Pathobiology* 74 (4): 206–11.

Haimes, Erica, and Michael Whong-Barr. 2004. "Key Issues in Genetic Epidemiology: Lessons from a UK Based Empirical Study." *TRAMES: A Journal of the Humanities & Social Sciences* 8.

Hastall, Matthias, Christoph Dockweiler, and Juliane Leinweber. 2017. "Achieving End User Acceptance: Building Blocks for an Evidence-Based User-Centered Framework for Health Technology Development and Assessment." 13–25. https://doi.org/10.1007/978-3-319-58700-4_2.

Hoffman, Sharona. 2021. "The Emerging Hazard of AI-Related Health Care Discrimination." *Hastings Center Report* 51 (1): 8–9.

Hoffman, Sharona, and Andy Podgurski. 2019. "Artificial Intelligence and Discrimination in Health Care." *Yale Journal of Health Policy, Law, and Ethics* 19: 1.

Holzinger, Andreas, Benjamin Haibe-Kains, and Igor Jurisica. 2019. "Why Imaging Data Alone Is Not Enough: AI-Based Integration of Imaging, Omics, and Clinical Data."

European Journal of Nuclear Medicine and Molecular Imaging 46 (13): 2722–30. https://doi.org/10.1007/s00259-019-04382-9.

Holzinger, Andreas, Georg Langs, Helmut Denk, Kurt Zatloukal, and Heimo Müller. 2019. "Causability and Explainability of Artificial Intelligence in Medicine." *Wiley Interdisciplinary Reviews: Data Mining and Knowledge Discovery* 9 (4): e1312.

Holzinger, Andreas, Carsten Röcker, and Martina Ziefle. 2015. "From Smart Health to Smart Hospitals." *Smart Health: Open Problems and Future Challenges*: 1–20.

Igbe, Michael A., and Clement A. Adebamowo. 2012. "Qualitative Study of Knowledge and Attitudes to Biobanking Among Lay Persons in Nigeria." *BMC Medical Ethics* 13: 1–9.

Institute of Electrical and Electronics Engineers. 2017. *Ethically Aligned Design: A Vision for Prioritizing Human Well-being with Autonomous and Intelligent Systems*. Version 2. https://standards.ieee.org/content/dam/ieee-standards/standards/web/documents/other/ead_v2.pdf (12.12.2024).

Jazieh, Abdul Rahman, and Zisis Kozlakidis. 2020. "Healthcare Transformation in the Post-Coronavirus Pandemic Era." *Frontiers in Medicine* 7: 429.

Jongsma, Karin R., and Annelien L. Bredenoord. 2020. "Ethics Parallel Research: An Approach for (Early) Ethical Guidance of Biomedical Innovation." *BMC Medical Ethics* 21 (1): 1–9.

Keane, Pearse A., and Eric J. Topol. 2021. "AI-Facilitated Health Care Requires Education of Clinicians." *Lancet* 397 (10281): 1254. https://doi.org/10.1016/S0140-6736(21)00722-4.

Keyes, Katherine M., and Daniel Westreich. 2019. "UK Biobank, Big Data, and the Consequences of Non-Representativeness." *Lancet* 393 (10178): 1297.

Khatib, Faisal, Dayana Jibrin, Joud Al-Majali, Mira Elhussieni, Sharifeh Almasaid, and Mamoun Ahram. 2021. "Views of University Students in Jordan towards Biobanking." *BMC Medical Ethics* 22 (1): 1–10.

Knapič, Samanta, Avleen Malhi, Rohit Saluja, and Kary Främling. 2021. "Explainable Artificial Intelligence for Human Decision Support System in the Medical Domain." *Machine Learning and Knowledge Extraction* 3 (3): 740–70.

Kozlakidis, Zisis. 2016. "Biobanking with Big Data: A Need for Developing 'Big Data Metrics'." *Biopreservation and Biobanking* 14 (5): 450–1.

Kozlakidis, Zisis. 2020. "Biobanks and Biobank-Based Artificial Intelligence (AI) Implementation Through an International Lens." In *Artificial Intelligence and Machine Learning for Digital Pathology: State-of-the-Art and Future Challenges*, edited by Andreas Holzinger, Randy Goebel, Michael Mengel and Heimo Müller, 195–203. Cham: Springer International Publishing.

Laal, Marjan. 2012. "Innovation and Medicine." *Procedia Technology* 1: 469–73.

Larson, D. B., et al. 2020. "Ethics of Using and Sharing Clinical Imaging Data for Artificial Intelligence: A Proposed Framework." *Radiology* 295 (3): 675–82.

Last, Mark, Olga Tosas, Tiziano Gallo Cassarino, Zisis Kozlakidis, and Jonathan Edgeworth. 2016. "Evolving Classification of Intensive Care Patients from Event Data." *Artificial Intelligence in Medicine* 69: 22–32.

Lee, Y., S. Roh, S. Donahue, K. Lee, and S. Kim. 2018. "Technology Acceptance Among American Indian Older Adults." *Innovation in Aging* 2 (Suppl 1): 169–70.

Lucey, Brendan P., Walter A. Nelson-Rees, and Grover M. Hutchins. 2009. "Henrietta Lacks, HeLa Cells, and Cell Culture Contamination." *Archives of Pathology & Laboratory Medicine* 133 (9): 1463–7.

Markus, Aniek F., Jan A. Kors, and Peter R. Rijnbeek. 2021. "The Role of Explainability in Creating Trustworthy Artificial Intelligence for Health Care: A Comprehensive Survey of the Terminology, Design Choices, and Evaluation Strategies." *Journal of Biomedical Informatics* 113: 103655.

Meacham, Sofia, Georgia Isaac, Detlef Nauck, and Botond Virginas. 2019. "Towards Explainable AI: Design and Development for Explanation of Machine Learning Predictions for a Patient Readmittance Medical Application." In *Intelligent Computing:*

Proceedings of the 2019 Computing Conference, Vol. 1, 939–55. Springer International Publishing.

Mittra, James, and Joyce Tait. 2012. "Analysing Stratified Medicine Business Models and Value Systems: Innovation-Regulation Interactions." *New Biotechnology* 29 (6): 709–19.

Murdoch, Travis B., and Allan S. Detsky. 2013. "The Inevitable Application of Big Data to Health Care." *Jama* 309 (13): 1351–2.

Nachmias, Rafi, and Marcia C. Linn. 1987. "Evaluations of Science Laboratory Data: The Role of Computer-Presented Information." *Journal of Research in Science Teaching* 24 (5): 491–506.

Naudé, Wim, and Nicola Dimitri. 2020. "The Race for an Artificial General Intelligence: Implications for Public Policy." *AI & Society* 35: 367–79.

Nemati, Shamim, Andre Holder, Fereshteh Razmi, Matthew D. Stanley, Gari D. Clifford, and Timothy G. Buchman. 2018. "An Interpretable Machine Learning Model for Accurate Prediction of Sepsis in the ICU." *Critical Care Medicine* 46 (4): 547–53. https://doi.org/10.1097/ccm.0000000000002936.

NIH Human Microbiome Portfolio Analysis Team. 2019. "A Review of 10 Years of Human Microbiome Research Activities at the US National Institutes of Health, Fiscal Years 2007–2016." *Microbiome* 7 (31). https://doi.org/10.1186/s40168-019-0620-y.

Noor, Poppy. 2020. "Can We Trust AI Not to Further Embed Racial Bias and Prejudice?" *BMJ* 368: m363. https://doi.org/10.1136/bmj.m363.

O'Sullivan, Shane, Andreas Holzinger, Kurt Zatloukal, Paulo Saldiva, Mohammed Imran Sajid, and Dominic Wichmann. 2017. "Machine Learning Enhanced Virtual Autopsy." *Autopsy & Case Reports* 7 (4): 3.

Owoyemi, Ayomide, Joshua Owoyemi, Adenekan Osiyemi, and Andy Boyd. 2020. "Artificial Intelligence for Healthcare in Africa." *Frontiers in Digital Health* 2: 6.

Pardes, Adam, William Lynch, Matthew Miclette, Ellen McGeoch, and Brian P. Daly. 2022. "Use of a Mobile Health (mHealth) Platform for Remote Assessment of Suicidal Ideation, Depression, and Anxiety: A Longitudinal Retrospective Study." *Innovations in Digital Health Diagnostics and Biomarkers* 2 (2022): 8–15.

Patrzyk, Piotr M., Daniela Link, and Julian N. Marewski. 2017. "Human-Like Machines: Transparency and Comprehensibility [Commentary]." *Behavioral and Brain Sciences* 40.

Paul, Shonali, Aditi Gade, and Sumani Mallipeddi. 2017. "The State of Cloud-Based Biospecimen and Biobank Data Management Tools." *Biopreservation and Biobanking* 15 (2): 169–72.

Raivola, Vera, Karoliina Snell, Ilpo Helén, and Jukka Partanen. 2019. "Attitudes of Blood Donors to Their Sample and Data Donation for Biobanking." *European Journal of Human Genetics* 27 (11): 1659–67.

Rajkomar, Alvin, Jeffrey Dean, and Isaac Kohane. 2019. "Machine Learning in Medicine." *New England Journal of Medicine* 380 (14): 1347–58.

Reis, Lea, Christian Maier, Jens Mattke, Marcus Creutzenberg, and Tim Weitzel. 2020. "Addressing User Resistance Would Have Prevented a Healthcare AI Project Failure." *MIS Quarterly Executive* 19 (4).

Reisman, Miriam. 2017. "EHRs: The Challenge of Making Electronic Data Usable and Interoperable." *Pharmacy and Therapeutics* 42 (9): 572.

Rivron, Nicolas, Martin Pera, Janet Rossant, et al. 2018. "Debate Ethics of Embryo Models from Stem Cells." *Nature* 564 (7735): 183–5.

Roberts, Huw, Josh Cowls, Jessica Morley, Mariarosaria Taddeo, Vincent Wang, and Luciano Floridi. 2021. "The Chinese Approach to Artificial Intelligence: An Analysis of Policy, Ethics, and Regulation." *AI & Society* 36: 59–77.

Robertson, John A. 2003. "Ethical and Legal Issues in Genetic Biobanking." In *Populations and Genetics*, 297–309. Leiden: Brill Nijhoff.

Rogers, Wendy Anne. 2004. "Evidence Based Medicine and Justice: A Framework for Looking at the Impact of EBM Upon Vulnerable or Disadvantaged Groups." *Journal of Medical Ethics* 30 (2): 141–5.

Rosemann, Achim, and Xinqing Zhang. 2022. "Exploring the Social, Ethical, Legal, and Responsibility Dimensions of Artificial Intelligence for Health-A New Column in Intelligent Medicine." *Intelligent Medicine* 2 (2): 103–9.

Roses, Allen D. 2000. "Pharmacogenetics and the Practice of Medicine." *Nature* 405 (6788): 857–65.

Rothstein, Mark A. 2005. "Expanding the Ethical Analysis of Biobanks." *Journal of Law, Medicine & Ethics* 33 (1): 89–101.

Sachs, Norman, Joep de Ligt, Oded Kopper, et al. 2018. "A Living Biobank of Breast Cancer Organoids Captures Disease Heterogeneity." *Cell* 172 (1): 373–86.

Sanderson, Saskia C., Kyle B. Brothers, Nathaniel D. Mercaldo, et al. 2017. "Public Attitudes Toward Consent and Data Sharing in Biobank Research: A Large Multi-site Experimental Survey in the US." *The American Journal of Human Genetics* 100 (3): 414–27.

Schneider, Johann Wilhelm, Micheline Sanderson, Dieter Geiger, Mostafa Nokta, and Sylvia Silver. 2016. "A Biobank to Support HIV Malignancy Research for Sub-Saharan Africa." *South African Medical Journal* 106 (9): 867–9.

Schreidah, Celine M., George Bingham Reynolds, Lauren M. Fahmy, et al. 2023. "Scoping Review of Genetic Databases for Rare Dermatologic Diseases: Opportunity for Artificial Intelligence and Machine Learning." *JAAD International* 12. https://doi.org/10.1016/j.jdin.2023.02.017.

Shaw, Brian. 1985. "The Role of the Interaction Between the User and the Manufacturer in Medical Equipment Innovation." *R&D Management* 15 (4): 283–92.

Singh, Hardeep, Christiane Spitzmueller, Nancy J. Petersen, Mona K. Sawhney, and Dean F. Sittig. 2013. "Information Overload and Missed Test Results in Electronic Health Record–Based Settings." *JAMA Internal Medicine* 173 (8): 702–4.

Song, Huan, Deepta Rajan, Jayaraman Thiagarajan, and Andreas Spanias. 2018. "Attend and Diagnose: Clinical Time Series Analysis Using Attention Models." *Proceedings of the AAAI Conference on Artificial Intelligence* 32 (1).

Sorell, Tom, Nasir Rajpoot, and Clare Verrill. 2022. "Ethical Issues in Computational Pathology." *Journal of Medical Ethics* 48 (4): 278–84.

Spruill, Ida J., Youlanda C. Gibbs, Marilyn Laken, and Tiffany Williams. 2014. "Perceptions toward Establishing a Biobank and Clinical Data Warehouse: Voices from the Community." *Clinical Nursing Studies* 2 (3): 97–106.

Thiel, Daniel B., Tevah Platt, Jodyn Platt, Susan B. King, and Sharon L. R. Kardia. 2014. "Community Perspectives on Public Health Biobanking: An Analysis of Community Meetings on the Michigan BioTrust for Health." *Journal of Community Genetics* 5: 125–38.

Thorogood, Adrian. 2020. "International Data Sharing and Rare Disease: The Importance of Ethics and Patient Involvement." *Rare Diseases. InTech Open*. https://doi.org/10.5772/intechopen.91237.

Thurow, Kerstin, et al. 2004. "Laboratory Information Management Systems for Life Science Applications." *Organic Process Research & Development* 8 (6): 970–82.

Tindana, Paulina, Cornelius Depuur, Jantina de Vries, Janet Seeley, and Michael Parker. 2020. "Informed Consent in Genomic Research and Biobanking: Taking Feedback of Findings Seriously." *Global Bioethics* 31 (1): 200–15.

Topol, Eric J. 2019. "High-Performance Medicine: The Convergence of Human and Artificial Intelligence." *Nature Medicine* 25 (1): 44–56.

Van Assche, Kristof, Serge Gutwirth, and Sigrid Sterckx. 2013. "Protecting Dignitary Interests of Biobank Research Participants: Lessons from Havasupai Tribe v Arizona Board of Regents." *Law, Innovation and Technology* 5 (1): 54–84.

Vandenberg, Olivier, et al. 2018. "Control of Infectious Diseases in the Era of European Clinical Microbiology Laboratory Consolidation: New Challenges and Opportunities for the Patient and for Public Health Surveillance." *Frontiers in Medicine* 5: 15.

Van de Poel, Ibo. 2016. "A Coherentist View on the Relation between Social Acceptance and Moral Acceptability of Technology." *Philosophy of Technology after the Empirical Turn*: 177–93.

Varzakas, Theodoros H., Ioannis S. Arvanitoyannis, and Haralambos Baltas. 2007. "The Politics and Science Behind GMO Acceptance." *Critical Reviews in Food Science and Nutrition* 47 (4): 335–61.

Vayena, Effy, Joan Dzenowagis, John S. Brownstein, and Aziz Sheikh. 2018. "Policy Implications of Big Data in the Health Sector." *Bulletin of the World Health Organization* 96 (1): 66.

von Wartburg, Walter P., and Julian Liew. 1999. *Gene Technology and Social Acceptance*. Lanham: University Press of America.

Vu, Hong Tien, and Jeongsub Lim. 2021. "Effects of Country and Individual Factors on Public Acceptance of Artificial Intelligence and Robotics Technologies: A Multilevel SEM Analysis of 28-Country Survey Data." *Behaviour & Information Technology*: 1–14.

Wolff, Justus, Josch Pauling, Andreas Keck, and Jan Baumbach. 2020. "The Economic Impact of Artificial Intelligence in Health Care: Systematic Review." *Journal of Medical Internet Research* 22 (2): e16866.

Wolsink, Maarten. 2012. "Wind Power: Basic Challenge Concerning Social Acceptance." *Encyclopedia of Sustainability Science and Technology* 17: 12218–54.

World Health Organization (WHO). 2021a. "Development of Guidance on Ethics and Governance of Artificial Intelligence for Health." *Consultation*. Geneva, Switzerland.

World Health Organization (WHO). 2021b. "Ethics and Governance of Artificial Intelligence for Health: WHO Guidance."

Wüstenhagen, Rolf, Maarten Wolsink, and Mary Jean Bürer. 2007. "Social Acceptance of Renewable Energy Innovation: An Introduction to the Concept." *Energy Policy* 35 (5): 2683–91.

Yala, Adam, Constance Lehman, Tal Schuster, Tally Portnoi, and Regina Barzilay. 2019. "A Deep Learning Mammography-Based Model for Improved Breast Cancer Risk Prediction." *Radiology* 292 (1): 60–6.

Yang, Qian, John Zimmerman, and Aaron Steinfeld. 2015. "Review of Medical Decision Support Tools: Emerging Opportunity for Interaction Design." *IASDR 2015 Interplay Proceedings*. https://www.researchgate.net/profile/Qian-Yang-19/publication/282121061_Review_of_Medical_Decision_Support_Tools_Emerging_Opportunity_for_Interaction_Design/links/56035cf608ae08d4f1716009/Review-of-Medical-Decision-Support-Tools-Emerging-Opportunity-for-.

Yang Meier, Dong, Petra Barthelmess, Wei Sun, and Florian Liberatore. 2020. "Wearable Technology Acceptance in Health Care Based on National Culture Differences: Cross-Country Analysis between Chinese and Swiss Consumers." *Journal of Medical Internet Research* 22 (10): e18801.

Yildirim, Nur, et al. 2021. "Technical Feasibility, Financial Viability, and Clinician Acceptance: On the Many Challenges to AI in Clinical Practice." In *Proceedings of the CHI Conference on Human Factors in Computing Systems (CHI '24)*, May 11–16, 2024, Honolulu, HI, USA. ACM, New York. https://doi.org/10.1145/3613904.3641896.

Zhang, Dongyu, Jidapa Thadajarassiri, Cansu Sen, and Elke Rundensteiner. 2020. "Time-aware Transformer-Based Network for Clinical Notes Series Prediction." In *Proceedings of the 5th Machine Learning for Healthcare Conference*. PMLR 126: 566–88.

10 Biobank-Based Research Employing AI Techniques

Challenges for Research Ethics Committees

Signe Mežinska

1 Introduction

Recent technological developments have opened new opportunities for employing artificial intelligence (AI) tools in data analysis for biobank-based research and for developing new AI tools for diagnostic, preventive, therapeutic, and other tasks in medicine based on biobank data. Biobanks are becoming crucial for this type of research, because they can provide the necessary amounts and types of data for AI-powered research studies and handle high-throughput sample and data operations (Narita et al. 2021; Kozlakidis 2020). Biobanks are also expected to play an important role in the increasingly complex governance frameworks that accompany development of AI and new technologies in general (Gille et al. 2020; Knight et al. 2024). An essential part of this governance process is the ethical review of biobank-based research studies. Given that biobank donors usually provide broad or open consent for storing their data and samples in biobanks and their use in research, which does not include informed consent for specific research studies, the ethics review process is paramount to ensure responsible use of samples and data and trustworthiness of biobanks. The introduction of new technologies, such as AI, in biobank-based research often raises novel ethical issues and poses challenges to the ethical review process implemented by research ethics committees (RECs) (Stahl et al. 2025).

RECs are a crucial element of the research ecosystem in academia, and there is growing pressure to establish ethics review bodies within the private sector as well; however, varying incentive systems and governance schemes in academia and the private sector may lead to regulatory frictions and discrepancies in application of ethical standards and review processes (Knight et al. 2024; Stahl et al. 2025). The primary task of RECs is to provide independent review to protect research participants' rights, evaluate the risk/benefit ratio, and maximize public benefit. The World Medical Association (WMA) Declaration of Taipei on Research on Health Databases, Big Data, and Biobanks defines the role of RECs in ethical review of biobank-based research as follows:

> the ethics committee must approve use of data and biological material and check whether the consent given at the time of collection is sufficient for the

DOI: 10.4324/9781032634487-15

> planned use or if other measures have to be taken to protect the donor. The committee must have the right to monitor ongoing activities.
>
> (WMA 2016)

RECs typically fulfill this role by assessing documents submitted by researchers, including study protocols, informed consent forms, and data management plans, before the research study begins to ensure ethical standards are met and potential risks to participants are minimized. Although the Declaration of Taipei mentions monitoring ongoing activities as one of the tasks of RECs, this function is rarely implemented outside the field of clinical drug research.

The effectiveness of RECs in fulfilling these tasks may be compromised by various factors, including insufficient financial and human resources, which can hinder the involvement of REC members and experts with adequate expertise and proper training, as well as influence execution of RECs functions (Coleman and Bouësseau 2008; Knight et al. 2024). REC members are typically chosen based on their diverse backgrounds to ensure a comprehensive review of research proposals, and they are individuals with expertise in various fields such as medicine, biology, ethics, social sciences, and law, as well as lay members with their unique views and experience. Nevertheless, evaluating biobank-based research that employs or develops AI systems often requires expertise from additional fields, such as computer science, data science, bioinformatics, specific legal expertise, and others which may not be well represented in RECs. Involvement of members with diverse scientific, ethical, and legal backgrounds is crucial to ensure a comprehensive review process. Additionally, independence of RECs and their individual members, along with the declaration and management of conflicts of interest is paramount to maintaining the trustworthiness of RECs (Lõuk 2023).

An analysis conducted by the SIENNA project concluded that the primary challenges faced by REC members in reviewing research that uses or develops AI tools include insufficient guidance and resources, limited expertise regarding the emerging ethical, legal, and social implications of AI (e.g., the impact of automated decision-making systems), lack of training for REC members, and the unpredictable development of AI technologies (Tambornino et al. 2019). Similarly, the Ada Lovelace Institute in its comprehensive 2022 report identified six challenges encountered by RECs when assessing the ethical implications of AI research:

1. Lack of the necessary resources, expertise, and training to adequately address the risks posed by AI.
2. Traditional research ethics principles are often not well-suited for the unique challenges presented by AI.
3. Ethical principles specific to AI-driven research are still developing and have not been consistently adopted by RECs.
4. Collaborations across multiple sites or between public and private entities can intensify existing challenges related to governance and decision-making.
5. RECs face difficulties in assessing the potential harms and impacts that may emerge throughout the entire life cycle of AI.

6. Private-sector RECs lack transparency in their processes, making it difficult to ensure that ethical standards are met.

(ALI 2022)

Also, the existing procedures and guidelines used by RECs may not sufficiently address the complexity and novelty of AI. Additionally, the current guidelines usually lack provisions for the continuous ethics monitoring of AI system development, which is crucial given the iterative training processes of many AI models.

Despite efforts to develop and disseminate guidelines and principles for ethical AI to various audiences, there has been comparatively less focus on integrating these principles into RECs' practices. In recent years, numerous documents from government bodies, scientific communities, industry leaders, and nonprofit organizations have articulated high-level AI ethics principles. However, these documents typically do not include specific guidance for RECs. For example, the Ethics Guidelines for Trustworthy AI by the Independent High-Level Expert Group on Artificial Intelligence (2019) emphasize the importance of ethical governance and accountability in AI research and development which aligns with the roles and responsibilities typically associated with RECs, such as ensuring ethical review and oversight, managing conflicts of interest, and maintaining trust in the research process. Nevertheless, the Ethics Guidelines for Trustworthy AI do not explicitly mention the role of RECs. Instead, these guidelines suggest the potential establishment of an "ethical AI review board" or a similar body to address overall accountability and ethical practices (EC 2019).

There are available some specific guidelines for RECs addressing the review of research involving AI, for example, Statement on Research Ethics in Artificial Intelligence by the Norwegian National Research Ethics Committees (NNREC 2019) and Research Ethics for AI Research Projects by the Centre Responsible Digitality (CRD 2023); however, these guidelines typically do not address specific issues distinctive to biobank-based research using AI. For example, in the context of REC reviews for biobank research, there are specific challenges related to informed consent when dealing with AI-driven studies. The existing broad and open consent typically applied for biobank-based research often does not adequately address the use of AI, and in many cases, it was obtained before the AI "revolution". Consequently, consent forms may lack specific information about how AI technologies may be used to analyze participants' data or how the data can be used for AI training (Andreotta et al. 2022; Kargl et al. 2022). This necessitates amending biobank consent forms and procedures to include steps for further communication and updates to participants about how their data is being used.

These observations of REC processes coincide with a rise in both the misuse of AI and accidental errors in its application. According to the AI, Algorithmic and Automation Incidents and Controversies (AIAAIC) database tracking cases of ethical misuse of AI, the number of incidents has increased 26 times since 2012 (Maslej et al. 2023) strongly demonstrating a need for strengthening ethics review and oversight.

This chapter analyzes the new challenges posed to the ethics review process by employing AI in biobank-based research. Examples of these challenges include epistemic challenges and their associated ethical implication (e.g., ethical aspects of AI knowledge production, understanding of novel risks, potential bias, technological reliability, epistemic justice, and transparency), as well as challenges in implementing research ethics principles (e.g., evaluation of risk/benefit ratio, protection of confidentiality and privacy, informed consent process, ensuring public benefit, and building trust). The aim of the chapter is to reflect on how to improve the REC review process for biobank-based research employing and developing AI techniques and identify the needs of RECs and other stakeholders in this context.

2 Epistemic Challenges in the Context of AI Use and Their Ethical Implications

Epistemic challenges refer to obstacles in the processes of generating and verifying scientific knowledge. In the biobank context, these challenges encompass issues related to standardization and harmonization of data to ensure that the data is "re-usable and re-purposable by bioscience researchers who did not create them" (Demir and Murtagh 2013), as well as broader questions about how knowledge is generated by using samples and data stored in biobanks, whose perspectives are included or excluded, and whether the knowledge produced is reliable and reproducible.

It is widely acknowledged that using incomplete or nonrepresentative datasets to train AI systems can result in different types of biases. This can also pose challenges to the robustness, accuracy, reliability, and generalizability of AI models, complicating the validation of findings (Schwartz et al. 2022; Eke and Shuib 2024; Adams 2024). These risks are amplified by the often complex and opaque nature of AI and machine learning, which can lead to "epistemic wrong" – "an injustice suffered by an agent in their capacity as a knower" – and "epistemic harm" – making "an agent qua knower worse off" (Dunne and Kotsonis 2024). So, in the biobank context, AI can potentially pose risks or even lead to negative consequences affecting how individuals and groups produce and use biomedical knowledge. Furthermore, epistemic harm may be compounded by other types of harm, such as discrimination or prejudice in different social and cultural settings, thereby intensifying the negative impact on individuals. Epistemic wrongs and harms become an ethical concern (Dunne and Kotsonis 2024; Fleisher and Šešelja 2023) because they may lead to violation of principles of biomedical ethics or perpetuation of inequalities in healthcare, for example, through misinterpreted data, leading to incorrect treatment recommendations.

Many AI models operate as "black boxes" where their decision-making processes are not interpretable, making it difficult for RECs, biomedical researchers, and practitioners to validate the reliability of AI insights. This opacity may reduce trust in AI applications in medicine and pose ethical challenges for patient safety and autonomy (Eke and Shuib 2024). Moreover, the lack of transparency in AI decision-making processes can influence trustworthiness of researchers and biobanks, as well as the ability of healthcare professionals to effectively use these technologies (Eke and Shuib 2024). Ensuring that AI models are explainable and their decision-making processes are transparent is crucial for maintaining ethical standards in research and healthcare settings and fostering trust in AI applications. Addressing these epistemic challenges in the process of REC review involves ensuring that the research process is transparent and just, with a commitment to the fair treatment of all individuals and groups involved in the production of knowledge. The EU Ethics Guidelines for Trustworthy AI emphasize the importance of transparency in AI systems. These guidelines advocate for making AI use clear and understandable, which is crucial when dealing with biobank data (EC 2019). Traceability mechanisms are recommended to ensure that AI systems and their decisions can be tracked and explained, providing clarity on how the data is used. This approach aims to foster trust and accountability in AI-driven research by making AI use more understandable and accessible to all users.

Nevertheless, without specific guidance and training, ethics review procedures may fail to account for the influence of the epistemic challenges raised by AI use mentioned earlier. For example, it might be hard to consider and evaluate the dynamic nature of AI algorithms, which can evolve and change over time, potentially altering their behavior and outcomes and influencing the knowledge production. Similarly, it might be hard to evaluate whether AI systems follow the transparency and accountability requirements. In this context, the needs of RECs and other stakeholders include training, clear guidelines, and collaborative frameworks. RECs need access to up-to-date information about AI technologies and their ethical implications, as well as tools to assess potential biases in AI systems. Both RECs and researchers require guidance on best practices for transparency in the use of AI, particularly regarding how participants' data will be used and shared. Biobank administrators, policymakers, and community representatives also play crucial roles in shaping ethical research practices in AI use, opening the need for ongoing dialogue among all stakeholders involved in the biobank-based knowledge production process.

2.1 Informed Consent

Implementing research ethics principles in biobank-based research using AI faces various challenges. As previously mentioned, biobanks encounter challenges with informed consent in AI-driven research. It is important to consider that the initial collection of data and samples is usually separated from the planning of specific research studies, and data may be sourced from multiple biobanks.

Traditional broad and open consent forms used by biobanks often lack information relevant to AI applications. This gap exists because many biobank consent forms were created before the widespread adoption of AI, and therefore did not anticipate the specific considerations and implications of using AI technologies. The rapid evolution of AI and the use of big data have often outpaced updates to consent forms, creating a gap in informing biobank participants about how their data might be used (Andreotta et al. 2022; Kargl et al. 2022). There also might be a lack of knowledge about how AI processes data, making it challenging for biobankers and researchers to provide detailed and transparent information to donors. Some authors even argue that biobank consent forms may provide misleading information about the range of lawful uses of biobank data which can lead potential biobank donors to believe their data will only be used for narrowly defined research purposes, while the actual scope is much broader (Kaaya 2024).

Transparent communication about AI use with biobank participants is crucial (see chapter "Tackling AI Transparency Concerns in Biomedical Research: Bringing a Communication-Participatory Approach to the Conversation" by Leslie Salgado Arzuaga), but often challenging to achieve during the initial informed consent process. One potential solution is introducing dynamic consent meaning "personalised, online consent and communication platforms" (Budin-Ljøsne et al. 2017). According to Budin-Ljøsne et al. (2017), dynamic consent approach is applied to achieve two goals: facilitating the informed consent process and enabling continuous communication between researchers and biobank participants (see also chapter "Artificial Intelligence and Communication With Research Participants in the Process of Biobanking Human Biological Material" by Jarosław Sak and Jakub Pawlikowski). Another potential solution has been suggested by Barnes et al. (2024), who introduced the concept of "demonstrated consent" leveraging blockchain and generative AI to enhance transparency and participant understanding in biobanking, and to maintain participants' trust. This approach proposes associating each biobank sample and associated data with a unique non-fungible token on a blockchain which records and updates information about the data and sample uses. The use of generative AI may help to provide this information in an understandable and interactive way and ensure a more transparent consent process. Nevertheless, both these suggestions do not solve all problems because their implementation requires significant resources, including technological infrastructure, and potentially they can lead to consent fatigue or perceived overload of information in the long term. For short-term solutions, RECs can initiate the review of applicability of the existing biobank consent forms to new types of research and promote updating consent forms to reflect the advancements in AI and data usage practices and re-consent procedures, if necessary. Involving AI, data protection, communication, and other relevant experts as REC members or outside experts might help ensure that the biobank consent forms are comprehensive, cover AI use, and are up to date with current ethical standards.

2.2 Confidentiality and Privacy

Protecting confidentiality and privacy is another major ethical challenge, given that developing and using AI models often require vast datasets that include sensitive health information. AI challenges existing approaches to privacy protection by using large amounts of data without predefined purposes, as well as challenging data protection principles like proportionality and minimization (Corfmat et al. 2025). Additionally, AI technologies are often controlled by private commercial companies, raising additional concerns about use and control of data. Some authors highlight risks posed by private actors in the context of AI and health data and provide examples where public–private partnerships have resulted in inadequate privacy protections (Murdoch 2021). Also, ensuring secure data handling and preventing unauthorized access are essential, but this may become challenging given the multiple stakeholders involved in research and the global nature of many biobank-based and AI-powered projects. Privacy protecting techniques, such as differential privacy (Subramanian 2023), federated access to data (Rujano et al. 2024), and others are promising but at the same time they may add complexity to data processing and research design and still struggle to provide full protection.

Even when biobank data are anonymized before being provided to the researchers, there is potential to reidentify individuals through complex data patterns and by connecting different data sources, which can threaten the privacy of biobank participants. Na et al. (2018) in their study have demonstrated an example where researchers using machine learning were able to reidentify more than 80% of individuals in a supposedly anonymized nationwide accelerometer-measured physical activity dataset by cross-referencing it with publicly available information. An additional concern for genetic data is genomic identifiability which has been demonstrated by the case of the HeLa cell line derived from the cervical cancer cells of Henrietta Lacks in 1951. These cells have been extensively used in scientific research and their reidentifiability has raised ethical concerns, especially regarding the privacy and consent of the Lacks family (Akyüz et al. 2024). These and other examples of reidentification point to challenges in ensuring true anonymity in research data.

2.3 *Risk/Benefit Ratio*

When reviewing research applications using AI, there are several specific risks that need to be carefully considered and mitigated, including epistemic risks and challenges mentioned earlier. At the same time, REC members often lack the expertise needed to evaluate the ethical implications of technology research, including AI, as their composition primarily includes medical doctors and bioethicists rather than technology researchers or social scientists (Stahl et al. 2025). This focus on biomedical risks means RECs may not be well-equipped to address privacy risks and broader social and epistemological impacts associated with AI research. Consequently, there is a need for RECs to adapt their expertise and frameworks to better assess ethical risks in evolving research practices beyond the biomedical realm (Stahl et al. 2025).

While the ethical considerations are paramount, there are ongoing discussions about whether it is also the responsibility of RECs to assess the reliability and validity of AI models for their intended use. Ensuring technical robustness is often intertwined with ethical considerations, as unreliable AI models can lead to ethical issues. Therefore, RECs may indeed play a role in evaluating both the ethical and technical aspects of research projects involving AI. A major concern is the possibility of bias in data collection and analysis, which can lead to unfair or inaccurate results, as discussed earlier. There is also a risk of exacerbating existing inequalities, as those members of the public with less access to biobank participation or facing other barriers may be underrepresented in datasets used in biobank-based research. Another challenge for REC members is understanding and mitigating novel risks introduced by AI for biobank participants, ensuring a favorable risk/benefit ratio. To evaluate the use of AI technologies, RECs need detailed information about the specific AI technology being used and its influence on biobank participants, patients, and society in general. To collect this information, RECs might need to amend their application forms to include sections specifically addressing AI-related risks and benefits, as well as data privacy and security measures. It is also crucial to ensure proper dialogue on ethical challenges after the ethics review.

Another ethical challenge is the implementation of the principle of justice. AI systems in biomedical research must be designed and used to provide broad benefits and avoid health inequities and they have potential to increase global health equity by providing efficient and cost-effective healthcare solutions to historically underserved groups of patients (Dychiao et al. 2024). Trustworthiness of biobanks and AI-driven biobank-based research hinges on a public perception that these technologies are aimed toward improving health for all.

3 Next Steps for Ensuring High-Quality REC Reviews

RECs currently play an important role in the biobank-based research ecosystem by ensuring adherence to ethical principles, which is essential for maintaining trust in science and innovation, particularly in biobanking (Knight et al. 2024). To fulfill this role, RECs must operate at various levels, including institutional, regional, national, and international, while adhering to established guidelines and legal requirements (Lõuk 2023). One direction for the implementation of guidelines and regulations to ensure high-quality review for biobank-based research studies involving AI and protection of biobank donors' rights is the development of dedicated training programs for REC members focused on AI technologies and their use in biomedical research and specifically in biobanking. A recent survey among RECs' members by Stahl et al. (2025) shows a high need for training specifically on AI. Such training should be organized by academic institutions or other stakeholders with expertise in ethical and technical aspects of research projects involving AI, and should include both theoretical and practical components, covering key ethical challenges and case studies. A blended learning format, combining online modules with onsite training could provide flexibility and ensure broader participation. Bearing in mind the rapid development of AI, regular training might be needed to keep REC members informed about the latest developments in the field. By equipping REC members with the necessary knowledge and skills, there is potential to improve the decision-making process.

Another step involving the potential for improving RECs' procedures is adopting more deliberative practices in communication with researchers and other stakeholders, including representatives of biobanks and members of public. This involves engaging in an open dialogue among RECs, researchers, biobankers, and the general public to better understand the ethical implications of AI technologies and ensure that all concerns are thoroughly discussed. It is also important for RECs to actively engage with researchers and other stakeholders throughout the review process, helping to identify potential ethical issues early on and allowing for collaborative problem-solving (Hickey et al. 2022). Critics for a long time have argued that ethics review procedures, while safeguarding research participants' well-being, may hinder ethical research practice and academic freedom (Allen 2008); nevertheless, there are responses saying that the real challenge lies not in RECs wanting to minimize ethical risks but in the often limited opportunities for dialogue and engagement between researchers and RECs (Hickey et al. 2022). Researchers who submit applications to RECs can often be viewed as experts in their respective fields. These researchers typically have extensive knowledge and experience in their areas of study, which is why they are conducting research that requires ethical review. Their expertise is crucial for understanding the complexities and nuances of their research projects, including the ethical implications. Recognizing researchers as experts can also facilitate more productive and respectful dialogue between them and the RECs. This mutual respect could lead to more effective communication, better understanding of ethical concerns, and ultimately, more robust and ethically sound research practices. Biobank representatives could play a crucial role in this communication process by acting as intermediaries who facilitate communication between researchers and RECs. They might also provide valuable insights into the ethical considerations specific to biobanking, such as issues related to consent, data privacy, and the use of biological samples. Additionally, their involvement could promote a more comprehensive understanding of the research ecosystem and the biobank's role within it, leading to more informed and balanced REC decisions. This collaborative approach might help enhance the quality of ethics reviews and support the development of ethical research practices that respect both research participants' rights and the scientific process and reduce researchers' worries about bureaucracy. Additionally, interdisciplinary collaboration is essential to address the multifaceted nature of epistemic challenges in AI research.

The dynamic and evolving nature of AI requires continuous updates to ethical guidelines to ensure they remain relevant and effective. A dual question is a necessity for the creation of new guidelines for reviewing AI-based biobank research proposals. If developed, such guidelines should emphasize critical ethical considerations specifically in the context of biobank-based research, such as including information about AI use in the informed consent process or the evaluation of the social and ethical impact of AI applications in biobank-based research. On the other hand, the integration of AI in biomedical research necessitates a re-evaluation of the existing ethical frameworks to accommodate the unique challenges posed by AI. It might be reasonable to evaluate the possibilities to include biobank-related aspects when updating the existing guidelines instead of creating new guidelines to avoid multiplication of documents and creating confusion.

When observing the rapid changes that AI is bringing into the biobanking field, some experts suggest that there is even a need for dedicated ethics committees for evaluation of ethical aspects of biobanking (Lõuk 2023) or specifically for AI-powered research, for example, the European Commission high-level expert group on AI in the Assessment List for Trustworthy Artificial Intelligence mentions "AI Ethics Review Board or AI Ethics Committee" (Ala-Pietilä et al. 2020). It is possible that dedicated RECs for biobanking and/or AI-powered research might offer specialized oversight, but at the same time, they would face several challenges, for example, regulatory fragmentation, lack of interdisciplinary collaboration, as well as the need for additional resources to establish and maintain such new committees.

To address the ethical challenges posed by AI in biobank-based research, we suggest a checklist for RECs to ensure addressing challenges in biobank-based research employing AI techniques (Table 10.1). The checklist provides a practical tool to ensure comprehensive ethical oversight, covering key areas such as informed consent, privacy, and risk assessment. It may help RECs evaluate the

Table 10.1 Checklist for RECs for Addressing Challenges in Biobank-Based Research Employing AI Techniques

Informed Consent	• Have biobank consent forms been updated to include information on the use of AI in biobank-based research and to reflect the advancements in AI and data usage practices? • Are there mechanisms in place to recontact biobank participants to obtain additional consent for the use of AI in research, if needed?
Privacy and Confidentiality	• Are commercial companies involved in a biobank-based research study involving AI? If yes, does it raise additional concerns about use and control of data? • What is the risk of reidentifying biobank donors in a biobank-based research study involving AI through complex data patterns or merging different data sources?
Risk/Benefit Ratio	• Does the REC application form include sections or questions specifically addressing AI-related risks and benefits, as well as data privacy and security issues? • Does REC have access to detailed information about the specific AI technology being used and its influence on biobank participants, patients, and society in general? • Has the applicant included a method for auditing AI systems for bias and implementing corrective actions (if applicable)? • Are there options to invite external experts for AI-related proposals? Are experts available? Does REC have funds to remunerate experts?
Dialogue With Researchers and Biobanks	• Do researchers and biobank representatives have an opportunity to schedule meetings with REC to discuss the ethical implications of the AI technologies they use? • Is the REC providing clear and constructive feedback to researchers throughout the review process to facilitate collaborative problem-solving? • Is there a process in place for reporting protocol deviations or adverse events regarding AI use?
Training	• Is training on AI technologies and ethical aspects of their use in biobanking available for RECs' members and other stakeholders?

ethical implications of AI technologies and foster collaboration between stakeholders. The checklist also emphasizes the importance of continuous education and training for REC members to stay up to date on AI advancements.

4 Conclusion

The integration of AI techniques into biobank-based research presents both significant opportunities and ethical challenges. To improve the REC review process, it is essential to adopt a structured approach that addresses informed consent, privacy, risk assessment, communication among the stakeholders, and continuous education of REC members. By ensuring comprehensive ethical oversight and fostering interdisciplinary collaboration, RECs can help better navigate the complexities introduced by AI technologies to all stakeholders involved. This approach may not only enhance the ethical governance of biobank-based research and AI use but also ensure that the needs of RECs and other stakeholders are effectively met. The biobanking community and RECs must regularly re-evaluate their ethical standards to keep pace with advancements in AI, and update ethics guidelines and practices as AI technologies evolve.

Reference List

Adams, Jonathan. 2024. "Introducing the Ethical-Epistemic Matrix: A Principle-Based Tool for Evaluating Artificial Intelligence in Medicine." *AI and Ethics*: 1–9.

Akyüz, Kaya, Melanie Goisauf, Gauthier Chassang, Łukasz Kozera, Signe Mežinska, Olga Tzortzatou-Nanopoulou, and Michaela Th Mayrhofer. 2024. Post-Identifiability in Changing Sociotechnological Genomic Data Environments." *BioSocieties* 19 (2): 204–31. https://doi.org/10.1057/s41292-023-00299-7.

Ala-Pietilä, Pekka, Yann Bonnet, Urs Bergmann, Maria Bielikova, Cecilia Bonefeld-Dahl, Wilhelm Bauer, Loubna Bouarfa, Raja Chatila, Mark Coeckelbergh, and Virginia Dignum. 2020. *The Assessment List for Trustworthy Artificial Intelligence (ALTAI)*. European Commission. https://ec.europa.eu/newsroom/dae/document.cfm?doc_id=68342 (08.06.2025).

ALI, Ada Lovelace Institute. 2022. "Looking Before We Leap: Expanding Ethical Review Processes for AI and Data Science Research." https://www.adalovelaceinstitute.org/wp-content/uploads/2022/12/Ada-Lovelace-Institute-Looking-before-we-leap-Dec-2022.pdf (08.06.2025).

Allen, Gary. 2008. "Getting Beyond form Filling: The Role of Institutional Governance in Human Research Ethics." *Journal of Academic Ethics* 6 (2): 105–16. https://doi.org/10.1007/s10805-008-9057-9.

Andreotta, Adam J., Nin Kirkham, and Marco Rizzi. 2022. "AI, Big Data, and the Future of Consent." *AI & Society* 37 (4): 1715-1728. https://doi.org/10.1007/s00146-021-01262-5.

Barnes, Caspar, Mateo Riobo Aboy, Timo Minssen, Jemima Winifred Allen, Brian D. Earp, Julian Savulescu, and Sebastian Porsdam Mann. 2024. "Enabling Demonstrated Consent for Biobanking with Blockchain and Generative AI." *The American Journal of Bioethics*: 1–16.

Budin-Ljøsne, Isabelle, Harriet J. A. Teare, Jane Kaye, Stephan Beck, Heidi Beate Bentzen, Luciana Caenazzo, Clive Collett, Flavio D'Abramo, Heike Felzmann, Teresa Finlay, Muhammad Kassim Javaid, Erica Jones, Višnja Katić, Amy Simpson, and Deborah Mascalzoni. 2017. "Dynamic Consent: A Potential Solution to Some of the Challenges of Modern Biomedical Research." *BMC Medical Ethics* 18 (1): 4. https://doi.org/10.1186/s12910-016-0162-9.

Coleman, Carl H., and Marie-Charlotte Bouësseau. 2008. "How Do We Know That Research Ethics Committees Are Really Working? The Neglected Role of Outcomes Assessment in Research Ethics Review." *BMC Medical Ethics* 9: 1–7.

Corfmat, Maelenn, Joé T. Martineau, and Catherine Régis. 2025. "High-Reward, High-Risk Technologies? An Ethical and Legal Account of AI Development in Healthcare." *BMC Medical Ethics* 26 (1): 4.

CRD, Centre Responsible Digitality. 2023. "Research Ethics for AI Research Projects." https://zevedi.de/wp-content/uploads/2023/02/ZEVED_AI-Research-Ethics_web_2023.pdf.

Demir, Ipek, and Madeleine J. Murtagh. 2013. "Data Sharing Across Biobanks: Epistemic Values, Data Mutability and Data Incommensurability." *New Genetics and Society* 32 (4): 350–65.

Dunne, Gerry, and Alkis Kotsonis. 2024. "Carving at the Joints: Distinguishing Epistemic Wrongs from Epistemic Harms in Epistemic Injustice Contexts." *Episteme*: 1–14.

Dychiao, Robyn Gayle, Lama Nazer, Donald Mlombwa, and Leo Anthony Celi. 2024. Artificial Intelligence and Global Health Equity. *British Medical Journal* 387. https://doi.org/10.1136/bmj.q2194.

Eke, Christopher Ifeanyi, and Liyana Shuib. 2024. "The Role of Explainability and Transparency in Fostering Trust in AI Healthcare Systems: A Systematic Literature Review, Open Issues and Potential Solutions." *Neural Computing and Applications*: 1–36.

European Commission. Directorate-General for Communications Networks, Content and Technology. 2019. "Ethics Guidelines for Trustworthy AI." Publications Office. https://ec.europa.eu/newsroom/dae/document.cfm?doc_id=60419 (08.06.2025).

Fleisher, Will, and Dunja Šešelja. 2023. "Responsibility for Collective Epistemic Harms." *Philosophy of Science* 90 (1): 1–20.

Gille, Felix, Effy Vayena, and Alessandro Blasimme. 2020. "Future-Proofing Biobanks' Governance." *European Journal of Human Genetics* 28 (8): 989–96. https://doi.org/10.1038/s41431-020-0646-4.

Hickey, Andrew, Samantha Davis, Will Farmer, Julianna Dawidowicz, Clint Moloney, Andrea Lamont-Mills, Jess Carniel, Yosheen Pillay, David Akenson, Annette Brömdal, Richard Gehrmann, Dean Mills, Tracy Kolbe-Alexander, Tanya Machin, Suzanne Reich, Kim Southey, Lynda Crowley-Cyr, Taiji Watanabe, Josh Davenport, Rohit Hirani, Helena King, Roshini Perera, Lucy Williams, Kurt Timmins, Michael Thompson, Douglas Eacersall, and Jacinta Maxwell. 2022. "Beyond Criticism of Ethics Review Boards: Strategies for Engaging Research Communities and Enhancing Ethical Review Processes." *Journal of Academic Ethics* 20 (4): 549–67. https://doi.org/10.1007/s10805-021-09430-4.

Kaaya, Emmi. 2024. "Biobank Consent under the GDPR: Are Potential Sample Donors Informed about All Lawful Uses of Biobank Data?" *Medicine, Health Care and Philosophy* 27 (4): 567–77. https://doi.org/10.1007/s11019-024-10227-6.

Kargl, Michaela, Markus Plass, and Heimo Müller. 2022. "A Literature Review on Ethics for AI in Biomedical Research and Biobanking." *Yearbook of Medical Informatics* 31 (1): 152–60.

Knight, Simon, Antonette Shibani, and Nicole Vincent. 2024. "Ethical AI Governance: Mapping a Research Ecosystem." *AI and Ethics*: 1–22.

Kozlakidis, Zisis. 2020. "Biobanks and Biobank-Based Artificial Intelligence (AI) Implementation Through an International Lens." In *Artificial Intelligence and Machine Learning for Digital Pathology: State-of-the-Art and Future Challenges*, 195–203. Springer.

Lõuk, Kristi. 2023. "Relevance and Challenges of Ethics Committees." In *Ethics and Responsible Research and Innovation in Practice: The ETHNA System Project*, edited by Elsa González-Esteban, Ramón A. Feenstra and Luis M. Camarinha-Matos, 125–36. Cham: Springer Nature Switzerland.

Maslej, Nestor, Loredana Fattorini, Erik Brynjolfsson, John Etchemendy, Katrina Ligett, Terah Lyons, James Manyika, Helen Ngo, Juan Carlos Niebles, and Vanessa Parli. 2023. "Artificial Intelligence Index Report 2023." *arXiv preprint arXiv:2310.03715*.

Murdoch, Blake. 2021. "Privacy and Artificial Intelligence: Challenges for Protecting Health Information in a New Era." *BMC Medical Ethics* 22 (1): 122. https://doi.org/10.1186/s12910-021-00687-3.

Na, L., C. Yang, C. C. Lo, F. Zhao, Y. Fukuoka, and A. Aswani. 2018. "Feasibility of Reidentifying Individuals in Large National Physical Activity Data Sets from Which Protected Health Information Has Been Removed with Use of Machine Learning." *JAMA Netw Open* 1 (8): e186040. https://doi.org/10.1001/jamanetworkopen.2018.6040.

Narita, Akira, Masao Ueki, and Gen Tamiya. 2021. "Artificial Intelligence Powered Statistical Genetics in Biobanks." *Journal of Human Genetics* 66 (1): 61–5.

NNREC, Norwegian National Research Ethics Committees. 2019. "Statement on Research Ethics in Artificial Intelligence." www.forskningsetikk.no/globalassets/dokumenter/4-publikasjoner-som-pdf/statement-on-research-ethics-in-artificial-intelligence.pdf.

Rujano, Maria A., Jan-Willem Boiten, Christian Ohmann, Steve Canham, Sergio Contrino, Romain David, Jonathan Ewbank, Claudia Filippone, Claire Connellan, Ilse Custers, Rick van Nuland, Michaela Th Mayrhofer, Petr Holub, Eva García Álvarez, Emmanuel Bacry, Nigel Hughes, Mallory A. Freeberg, Birgit Schaffhauser, Harald Wagener, Alex Sánchez-Pla, Guido Bertolini, and Maria Panagiotopoulou. 2024. "Sharing Sensitive Data in Life Sciences: An Overview of Centralized and Federated Approaches." *Briefings in Bioinformatics* 25 (4). https://doi.org/10.1093/bib/bbae262.

Schwartz, Reva, Reva Schwartz, Apostol Vassilev, Kristen Greene, Lori Perine, Andrew Burt, and Patrick Hall. 2022. *Towards a Standard for Identifying and Managing Bias in Artificial Intelligence*. Vol. 3. US Department of Commerce, National Institute of Standards and Technology.

Stahl, Bernd Carsten, Etienne Aucouturier, Jurate Lekstutiene, Tom Lindemann, Maria Maia, Ana Marušić, Antonija Mijatović, Elahe Naserianhanzaei, George Ogoh, and Anais Resseguier. 2025. "Knowledge Needs of Research Ethics Committees for the Integration of Ethics in Research and Technology Development." *Journal of Responsible Innovation* 12 (1): 2466904.

Subramanian, Rishabh. 2023. "Have the Cake and Eat It Too: Differential Privacy Enables Privacy and Precise Analytics." *Journal of Big Data* 10 (1): 117. https://doi.org/10.1186/s40537-023-00712-9.

Tambornino, Lisa, Dirk Lanzerath, Rowena Rodrigues, and David Wright. 2019. *SIENNA D4. 3: Survey of REC Approaches and Codes for Artificial Intelligence & Robotics*. https://doi.org/10.5281/zenodo.4067990 (08.06.2025).

World Medical Association. 2016. "Declaration of Taipei on Ethical Considerations Regarding Health Databases and Biobanks." www.wma.net/policies-post/wma-declaration-of-taipei-on-ethical-considerations-regarding-health-databases-and-biobanks/ (30.01.2025).

11 Powerful AI

Between Accountability and Potentiality

Michaela Th. Mayrhofer

1 Introduction

In 2008, *Time Magazine* highlighted biobanks as one of the "ten ideas that are changing the world right now" (Park 2009). This perception continued into the next decade, with a *Forbes* article eventually proclaiming that thanks to the rapid ascension of artificial intelligence (AI) and its "transformative potential", "biobanking is here" (Pandya 2019). "About time", one is tempted to sigh. Biobanking, however, has been around for a long time and has steadily gained new significance and meaning (Mayrhofer 2013a). Put differently, biobanking is not a new practice and AI goes beyond being a simple technical tool for unearthing the untapped potential attributed to biobanks. The growing amount, variety, and accessibility of data – gathered for scientific discovery, lifestyle, disease prevention, and therapy – are driven by the belief that extensive datafication will enhance precision, ultimately leading to improved health outcomes through (open data) innovation. This is supported by systems that are designed for efficient automation, nurturing a narrative with an optimistic outlook on technological growth (Hoeyer 2019). Data, in this future, are framed as pure, innocent, raw, and all-powerful, in other words, "immaculately conceived" (Bronson 2022).

Collecting biological materials in both the medical and natural sciences dates to the eighteenth century, with a particular emphasis in the nineteenth century on satisfying scientific curiosity by distinguishing the normal from the pathological (Hilgartner 1995). Over the years, collections of biological samples and data have developed for archival, diagnostic, research, or therapeutic purposes, eventually increasing in scale as well as processing and service capacity. In the following centuries, the focus shifted to the cellular level (Landecker 2007), with continued importance placed on the materiality of the sample and its transformation into "biovalue" (Waldby 2002; Waldby and Mitchell 2006) that refers to the new meaning manufactured through biotechnological practices.

Biobanking is the practice of collecting, registering, processing, preserving, storing transforming, and distributing biological material and its associated data in a standardized setting (Cambon-Thomsen 2004; Mayrhofer 2013a). Knowledge, in the form of data or information, was primarily derived from the sample itself and served a supporting role while recognizing the core principles of archival

DOI: 10.4324/9781032634487-16

provenance. Information that described the sample in basic terms was later supplemented by additional details from medical records or scientific evidence, obtained through meticulous laboratory work. Simultaneously, biobanks were depicted as "gold mines" (Hagen and Carlstedt-Duke 2004) or "organic bank accounts" (Park 2009), in other words, as resources that only needed to be mined with the right tools. As their quality and quantity grew, biobanks reached a "critical mass" that revealed further potential and new weight of an old practice and now recognized resource that satisfies not only a well-defined archival or diagnostic purpose but also a wide variety of yet-to-be-defined therapeutic or research purposes (Mayrhofer 2013a).

Timothy Lenoir (2002) observed early on that the database began to challenge the laboratory as the primary site of knowledge production. Especially, as biobanks commenced to challenge the seemingly clear-cut distinction between the materiality of the sample and the information value of generated as well as (re)combined data. Concurrently, the data(base) itself became both a research tool and an "epistemic object" (Rheinberger 1997) – with raw data being filtered in datasets producing new knowledge, being dynamic, and continuously evolving over time. This so-called "informational turn" (Beaulieu 2004) of the biological sciences promises greater comparability, easier access, and enhanced data exchange, among other things.

Today, data are no longer just descriptions with varying degrees of provenance; they have become central objects and active agents, as significant as physical samples and, in some cases, even taking precedence over them, depending on the context. As a result, biobanks could today be redefined as collections of data and samples, with data (*in silicio*) potentially taking precedence, provided they are gathered in accordance with the necessary quality standards of data mining. Besides data quality, successful data usage requires robust tools and sound access conditions (technical as well as ethico-legal). High-quality data ensure that the insights generated are accurate and meaningful, minimizing the risks of drawing incorrect conclusions and ultimately being reproducible. Machine learning algorithms or software solutions, for instance, allow for efficient processing, pattern recognition, and predictive analytics. Finally, proper access procedures and policies ensure that data is managed accordingly and available.

Given the significant public funding and political endorsement for biobanks and research infrastructures, the underuse of data and biological material has been identified as a persistent problem (Rush et al. 2020; Cadigan et al. 2014; Beier and Lenk 2015; Mayrhofer 2013b; Oldoni et al. 2021; Hämäläinen et al. 2019). Likewise, as a precondition to fully harness the scientific as well as economic potential of large datasets, the digital objects must be (made) findable, accessible, interoperable, and reusable (FAIR). These FAIR principles were designed to provide guidance for improved data management and stewardship of quantitative and qualitative data, metadata, algorithms, or software across all sciences (Wilkinson et al. 2016). When it concerns highly sensitive data such as health data, it was argued that biological material and data should be considered as a unified resource to enable the integration of comprehensive provenance information, including data reuse, and to incentivize the repurposing of data and materials by using privacy-preserving methods (Holub et al. 2018), as well as to ensure the upholding of ethical, equitable, and responsible practices (Murtagh et al. 2021; Austin 2020) or assessing fairness (see chapter "Fairness of an AI System in the Case of a Biobank of Images and Imaging Biomarkers" by Valentina Colcelli).

Seeing that the associated and derived data extend beyond the materiality of the sample, I propose a more radical rethinking. Biobanks should be seen as collections of both, the body (social) and data. These two objects are interconnected, depending on the context, making it necessary to focus on the "and" rather than simply the "body (sample)" or "data". Furthermore, going beyond the definition of the biobank as a repository of samples "and" data, the "organism" metaphor allows to grasp a biobank as a growing, living organism within a particular environment (Akyuz et al. 2024). This metaphor also extends to governance aspects, integrating societal elements that shape the establishment and account for the continuous improvements necessary to maintain a biobank and keep it accessible as well as relevant. The *in vivo* element goes beyond the sample and includes not only the people who contribute to the biobank (e.g., research participants and funders) but also those who use its services (e.g., researchers), as well as those who translate policies, recommendations, and findings into routine procedures (e.g., biobank staff). These processes shape the design and evolution of the governance framework and ultimately form the "organism" called biobank. This also reflects what Kean Birch (2017) has termed "valuation practices", where political-economic processes manage what is not inherent in biological materialities alone but a constant, dynamic process of value generation.

Value generation is neither a neutral nor purely economic process; rather, it is deeply intertwined with power dynamics, as conceptualized in the elusive and compelling works of Michel Foucault. "Biopower", for instance, operates as both a disciplinary and a productive force, shaping individual behaviors while also structuring societal norms, institutions, and governance (Foucault and Rabinow 1984, 2002). Moreover, many feminist scholars, most notably Donna Haraway (1988), have pointed out that scientific claims are rooted in power all the way down, and that it is precisely this power that drives societal and political progress. AI systems function as tools of governance, exercising (bio)power through predictive analytics, data collection, and machine-based decision-making. To what end this productive power is used is yet to be negotiated, heavily depending on the societal values of a given society. Shoshana Zuboff (2020) argues that the commodification of personal data can lead to unprecedented power imbalances and a new economic order in knowledge production by actors in both the public and private sectors, which she coins "surveillance capitalism". Subsequently, the widespread collection of personal data, as Ashton Black (2025) argues, risks a deepening of the divide in knowledge production by reinforcing injustices, unless a raised public awareness about the wider societal impact is realized, ideally through civil society organizations and the media.

Consequently, in this chapter, I draw the attention to the complex relationship between power, potentiality, and accountability exemplified by the technology and governance of AI within the context of the practice of biobanking in Europe. As a key analytical concept, I further rely on the often uninterrogated "socio-technical imaginary" (Jasanoff and Kim 2015), which enables the examination of (infra) structure–agency relationships through interpretative research and analysis. It allows highlighting how narratives and framings of accountability and potentiality shape imaginaries and assign meaning to potential futures, especially when translated into scientific and governing practices of biobanking.

2 The Strategic Potential of AI

Biobanks, as infrastructures and service providers, are situated in the intersectoral space between healthcare and health research and are used to adapting new technologies and standards for efficient data management. Inevitably, they are subjected – one way or the other – to the widespread entrepreneurial confidence in AI systems.

Lately, it seems that new benchmarks for deep learning, AI-driven innovation, and large-scale investment announcements are topped every other day, while application areas are yet to be determined. Among the latest was the news about the Chinese AI startup DeepSeek, which sent ripples through global markets and triggered a tech-driven sell-off in the US and Europe (Teng 2025). The unexpected news furthermore caused investors and policymakers to reassess national strategies, regulations, and funding schemes. More than highlighting the increasing influence of Chinese AI firms on the global tech landscape, this emphasizes that the AI race is now openly political and impacts multiple sectors. Consider as case in point the timing and amount of recent investment pledges to topple investments, such as of $112 billion in France alone (Stargardter 2025).

The expected progressive potential of AI has driven many countries to invest substantial public and private funds, which should be integrated into cohesive digital health strategies. Whereas on a general level, the EU leans toward a regulation-focused approach, ensuring AI aligns with principles of accountability and fairness, the US emphasizes innovation and market-driven strategies, with relatively less regulation but significant focus on national security and maintaining technological dominance. China, in contrast, drives innovation and commercialization through government leadership, focusing on national security and surveillance.

All three (the US, China, and EU) have in common that the strategic potential of AI has been grasped at both the entrepreneurial as well as the political level. Consider, for instance, the recent announcement of a $500 billion initiative to establish the US dominance in AI, beginning with a massive data center in Texas and the backing by private funding and major partners such as SoftBank, OpenAI, Oracle, Nvidia, Microsoft, and Arm. In contrast, Europe has no comparable AI companies (yet), able to attract similar private investment (Haeck 2025). That said, however, the race is on – across all sectors.

Despite the acceleration of news about cheaper, better, and faster AI-driven solutions in recent days, the potential for health – especially in regard to efficiency – has been recognized for some time. The 2024 edition of Health at a Glance: Europe by the Organization for Economic Co-operation and Development (OECD 2024), for instance, identified that managing the shifting balance between demand and supply capacity in European health systems is a fundamental challenge for which digital transformation is key. Advocating for greater efficiency, digital tools are recognized for their potential to reduce the administrative workload of healthcare professionals by up to 30%, with the European Health Data Space (EHDS) identified as a crucial framework for data exchange and accessibility. Success in this digital transition requires strengthening health data infrastructure, incorporating comprehensive cybersecurity measures, and providing targeted training for both health workers and patients. Against this backdrop, the OECD continues to support countries in implementing the 2017 OECD Council Recommendation on Health Data Governance, including efforts focused on digital security in health and the responsible use of AI in healthcare (ibid, 4).

Automation is expected to improve efficiency, particularly in repetitive back-office procedures and AI-assisted initial screenings of medical images, which can accelerate diagnostic processes with greater accuracy in pattern recognition than humans (KPMG Enterprise 2018). Especially in the fields of radiology and pathology, AI is demonstrating a significant value gain in the interpretation of diagnostic images arguing with an efficiency narrative. For instance, researchers at Stanford have developed an algorithm capable of analyzing and labelling chest X-rays within a significantly shorter time frame:

> The average time for radiologists to complete labeling of 420 chest radiographs was 240 minutes (range 180–300 minutes). The deep learning algorithm labeled the same 420 chest radiographs in 1.5 minutes and produced heat maps highlighting areas of the image that are indicative of a particular pathology in 40 additional seconds.
>
> (Rajpurkar et al. 2018, 18)

Continuing with the example of chest radiographs, advancements in AI continue to support diagnostic processes, even though data imbalances (e.g., quality, shortage, and access) largely remain. Yet, suggesting to focus on data aspects and their technological exploitation alone (Myong et al. 2023) reinforces the narrative of "immaculate conception" (Bronson 2022) and optimistic futures due to technological advancements (Callebaut 2012; Mittelstadt and Floridi 2016; Hoeyer 2019). In other words, it disregards that "AI tools in healthcare can be prone to errors and patient harm, biases and increased health inequalities, lack of transparency and accountability, as well as data privacy and security breaches" (Lekadir et al. 2025).

3 About Core Values and National Strategies

Brent Mittelstadt (2021) has pointed out that digitalization and AI present both opportunities and risks. I read this, not in the sense of two sides of the same coin, where one necessarily follows or counterbalances the other, but rather as forces that unfold simultaneously, and are interwoven yet, at times, independent in their effects. The potential outcomes of digitalization and machine learning neither exist in isolation nor as mere trade-offs. Instead, they emerge in parallel, shaping and influencing society and societal values at the same time. While AI advancements can drive innovation, efficiency, and progress, they can also initiate ethical issues, biases, and unintended consequences, all developing concurrently rather than sequentially. Consequently, suboptimal implementation of technically advanced tools and misalignment in system updates, could, for instance, further burden the already overstretched clinical workforce. To mitigate such a risk, Mittelstadt argues for gradual training programs and well-planned implementation strategies. These should not only ensure that healthcare professionals can effectively use digital tools but also facilitate communication with colleagues and patients, ultimately reshaping the doctor–patient relationship in ways that are yet to be fully understood.

This is equally true for any sensitive data-fueled endeavor, whether it is clinical research or population biobanking. Accordingly, trustworthy AI in (bio)medicine and underlying values are increasingly becoming a topic for governance (specialists), at times even a regulatory and political priority, that still needs to be translated for daily routines and practices. International scholars explore normative concepts such as fairness, explainability, robustness, transparency, bias mitigation techniques, liability aspects, solidarity, or altruism, among others (Bak et al. 2023; see chapter "Fairness of AI system in the case of a biobank of images and imaging biomarkers" by Valentina Colcelli; Buyx and Prainsack 2018; de Ruijter et al. 2024; Fehr et al. 2024; Hummel and Braun 2020; Okaibedi Eke et al. 2025; Zhang and Zhang 2023), while arguing for applying these principles already while designing AI-driven technologies and applications (Lekadir et al. 2025; Müller et al. 2021). This is tricky in and for practice, as data-driven medicine exemplifies that the individual and collective good are intertwined. Especially, since the motivations for donating health data differ from person to person; whether it is to support research in general, to benefit directly, or to help maintain a solidarity-based healthcare system for the next generation (Hummel and Braun 2020). Consequently, institutions that are transparent about their data usage and access conditions, see it as their responsibility and mandate to generate social value in the first place (Prainsack 2017), in contrast to those who focus predominantly on economic gain.

Hence, the advancement and integration of AI in healthcare and health research involve a delicate balance of trade-offs, perhaps most prominently between the free-market dynamics and social systems. On one hand, there is a strong push to uphold stringent data privacy measures; on the other, fully unlocking AI's transformative potential for the better requires broad access to health data, as is the case with biobanks. These two goals are often at odds with each other when defining the specifics of access conditions. To navigate this challenge, countries must develop and implement digital health strategies that are not only coherent but also aligned with their fundamental values (human dignity, freedom, democracy, equality, rule of law, and human rights) and are reflected in their (trans)national data governance frameworks. This necessitates an open and transparent dialogue with citizens, ensuring that policy decisions regarding data privacy and accessibility are well-justified, trustworthy, and widely understood, especially in relation to the management of sensitive data. Governments must be able to articulate the reasoning behind their chosen priorities, reflected in their choices for funding allocations (Bak et al. 2022), particularly when a national strategy stresses accountability, transparency, and trustworthiness as key principles.

In 2017, according to the OECD, only a few countries had developed national AI strategies. In contrast, over 50 national strategic and government-wide initiatives for trustworthy AI have been recorded, along with more than 930 policy initiatives across 71 jurisdictions, since the adoption of the OECD AI Principles in 2019 (Russo and Oder 2023). Ethical and societal aspects should be(come) central to discussions on the future of AI in healthcare, ensuring that innovation benefits all members of society, rather than reinforcing inequalities, especially when there is a political commitment to ensure solidarity-based healthcare, where social protection and cohesion is depicted as a core value. At the same time, however, it remains unresolved whether data privacy or data solidarity should be a foundational value of research policies (Bak et al. 2023).

4 About Regulating the Potential of AI in Europe

Guiding principles, while not legally binding, are a form of soft law that serve as normative benchmarks. They are often endorsed (and noncompliance is sanctioned) by the respective peer groups, in some cases institutionalized or enforceable through funding requirements. A recent study across different sectors analyzed 200 policies and ethical guidelines by public bodies, academic institutions, the private sector, and nongovernmental organizations across the globe, identifying at least 17 groups of resonating common principles (Correa et al. 2023).

So far, the European Union (EU) has played a pivotal role in shaping global debates on data governance and has set a benchmark for privacy regulations worldwide, particularly through the General Data Protection Regulation (GDPR). With regard to research and biobanking, the GDPR provides a favorable while complex framework along with national laws and regulations, aiming to balance individuals' fundamental rights, business freedom (recital 4), and society's legitimate expectation for increased knowledge (recital 113). This balance is achieved by allowing research to proceed with fewer restrictions, if adequate safeguards are in place to protect individuals' rights and freedoms (Article 89.1) and providing specific exemptions (Slokenberga et al. 2021). Despite criticism of being over-regulative or leaving much regulatory fine-tuning possibilities to the member states (Ducato 2020), the European Data Protection Supervisor (EDPS) noted that there is no evidence to suggest that the GDPR actually hinders legitimate scientific research by design. It recommended, however, that Data Protection Authorities (DPAs), Research Ethics Committees (RECs) (see also chapter "Biobank-Based Research Employing AI Techniques: Challenges for Research Ethics Committees" by Signe Mežinska), and the research community engage in a closer dialogue (European Data Protection Supervisor 2020).

> The EU Ethics appraisal process considers 'ethics' as including questions of legal and regulatory compliance as well as questions on moral principles, fundamental rights and values, and how these apply to research activities. It is part of a process of 'governance', that aims to ensure the protection of the rights and interests of all those affected by research and related activities, respect for fundamental EU values and human rights, and that EU funded research is not misused.
>
> (European Commission 2023)

The European Commission can enforce compliance with soft law by requiring that all research and innovation activities funded under its Framework Programs, such as Horizon 2020, adhere to ethical principles as well as relevant national, EU, and international laws. This includes, among others, the Council of Europe's Convention for the Protection of Human Rights and Dignity of the Human Being with regard to the Application of Biology and Medicine: Convention on Human Rights and Biomedicine (Oviedo Convention) and additional protocols, the World Medical Association's Declaration of Helsinki (2024), the Charter of Fundamental Rights of the European Union (2012), or the European Convention on Human Rights, along with its supplementary protocols. In relation to integrating AI into the health sector, many governments, international bodies, and funding institutions recognize the need for clear ethical and legal frameworks to ensure patient safety, fairness, and transparency. In April 2021, the European Commission proposed the world's first-ever regulatory framework on AI, which famously employs a risk-based approach (see also chapter "A Proposal for a Combined Risk Management Approach to AI Applications in Health Research and Biobanking" by Ilaria Colussi), and entered into force as Artificial Intelligence Act in August 2024. The EU addresses AI as part of its wider digital strategy aiming for better conditions for its development, as well as balancing entrepreneurial and societal aspects. Moreover, the stated objective is to develop regulations that not only address local and regional concerns but also have a far-reaching global impact, shaping international standards and fostering global collaboration in the process. The ambition is to establish a gold standard for best practices and regulations that ensure emerging technologies, such as AI, are governed in a way that promotes safety, fairness, and accountability across borders. In the words of the European Commission, on the dedicated subpage on Excellence and Trust in Artificial Intelligence, it reads under the heading Benefits of AI:

> The EU has the potential to become the global leader in safe AI. By developing a strong regulatory framework based on human rights and fundamental values, the EU can develop an AI system that benefit[s] people, businesses and governments.
>
> (European Commission 2021a)

Looking at the EU and its regulation-driven approach, its funding for healthcare and medical AI-related research projects has increased through the EU's Horizon frameworks, alongside the establishment of several ethical and legal instruments, most notably the Artificial Intelligence Act or the Guidelines of Trustworthy AI (European Commission 2019). The latter defined early on seven equally critical key requirements that AI systems should meet to be worthy of the user's trust: (1) Upholding human agency, AI should empower humans to make informed decisions while ensuring proper oversight through approaches like human-in-the-loop and human-in-command. (2) Ensuring technical robustness and safety, AI must be secure, resilient, and reliable, with fallback plans to minimize unintentional harm. (3) Respecting privacy and data governance, AI requires data quality, integrity, and authorized access modalities. (4) AI systems, their data, and business models should be transparent, with traceability and explanations of decisions tailored to stakeholders. This includes to inform users when they are interacting with AI, understanding its capabilities and limitations. (5) A commitment to diversity, non-discrimination, and fairness should avoid bias and promote inclusivity. (6) For societal and environmental well-being, AI should be not only sustainable but benefit all humans and future generations. (7) Auditability and clear accountability mechanisms shall be put in place for AI systems, especially for critical applications.

In this chapter, I choose not to delve too much into the actual impact of guidelines, neither generally nor specifically for biobanks, but instead direct readers to a selection of relevant scholarly articles that explore this issue in depth (Correa et al. 2023; Goisauf and Cano Abadia 2022; Radclyffe et al. 2023; Ribeiro Junior 2024; Vetter et al. 2023). Rather than analyzing the guidelines' effectiveness, my focus here is to emphasize that political will and commitment to balancing social values are deeply embedded in strategic frameworks and policy initiatives. That said, I want to stress my agreement with one argument: that by placing the emphasis on the normative requirements of the AI system itself, one effectively narrows the ethical debate to the interaction with a particular application rather than the technology at large. This approach, while critical for compliance aspects, risks overlooking the broader sociopolitical, economic, and structural dimensions that shape the impact of this technology. It further jeopardizes sidelining discussions on accountability that are deeply embedded in power dynamics and require a comprehensive debate beyond mere compliance, but assess the situatedness and ecosystems in which these technologies are designed, implemented, and ultimately governed (Bak et al. 2020).

4.1 Negotiating Social Value: Interplay of Power and Accountability

The technological and industrial potential of emerging innovations is closely intertwined with the (re)negotiation of their social value. While such advancements promise significant benefits and are typically depicted positively (Callebaut 2012; Mittelstadt and Floridi 2016; Hoeyer 2019), their impact is not solely determined by technical capabilities, but equally shaped by expectations that are rooted in respective societal value systems. For biobanks, this requires constantly adapting to match the expectations of not only funders and regulators but also research participants and citizens, as well as researchers (users).

In general, AI-driven technologies and practices are typically framed through an optimistic lens, emphasizing progress while sometimes neglecting potential inequalities, biases, or ethical dilemmas. Therefore, it is crucial to critically examine not only the expected outcomes but also who benefits from them in the end. If a national strategy is drafted, for instance, with the ambition and promise of keeping a publicly funded healthcare system in operation, and data is collected from research participants and citizens under the same promise, then it is essential to ensure that technological progress aligns with the agreed-upon societal values of a society and its national priorities.

If this promise is not kept – such as when access to healthcare becomes exceedingly costly, or when research is not conducted as consented – the incentive to donate samples and data is lost, whether for altruistic, solidarity, return-on-investment, or personal reasons such as family health concerns. In other words, the social contract is broken.

Being situated at the intersection of healthcare and health research, the practice of biobanking provides a good case for examining how the implementation of emerging technologies such as AI can challenge the upholding and meaningful renegotiation of the social contract and its underlying core values. In general, new technologies are positively framed by and for biobanks, without neglecting the challenges. At the same time, the governing ethical and legal requirements are often depicted as obstacles to fully realizing the favorable potential of such technologies. Especially the GDPR is often criticized, either generally or in specific details (Peloquin et al. 2020; Becker et al. 2024). Although some criticism is justified and constructive, I argue that the challenge the GDPR imposes is less a legal and more a structural one, especially for the complex and fragmented social and technological systems that govern healthcare and health research.

Legal departments of public research institutions, for instance, are notoriously understaffed and tasked to primarily protect the integrity and liability of the organization, ensuring it is shielded from legal risks and reputational harm, rather than to strike the balance to enable research (see also chapter "Liability for Damages, AI, and Machine Learning for Digital Pathology" by Dorota Krekora-Zając). As a result, the ethical and societal implications are still often regarded as secondary concerns. They tend to receive insufficient attention or are depicted as mere bureaucratic necessities, rather than being rewarded and fostered as integral components of a trustworthy governance framework. Systemic challenges like these require a cultural shift and a change in mindset for which political will and acceptance by those most affected are preconditions for the meaningful and sustainable adaptation (see chapter "Social Acceptance of Artificial Intelligence and Biobanking" by Elodie Caboux et al.). Consequently, these solutions and their implications must be addressed both locally (e.g., at the clinic) and on a transnational level, as they concern staff training, career development, legal and ethical compliance, and policy implementation in line with the commitment of social values such as non-discrimination or human oversight.

Perhaps Effy Vayena (2021) put it most eloquently, when she highlighted that three challenges must be overcome before the benefits of data, AI, and digital health can be realized. This includes, firstly, technical and governance standards that must be harmonized – not only technically, by defining and ensuring interoperability, but also by guaranteeing the ethical use of data and the development of trustworthy digital health applications in tandem. Secondly, since pilot AI studies provide only limited insights into the impact of heterogeneous real-world care (and, I argue, research) settings, the evidence of their impact needs to be assessed systematically and interdisciplinarity. Thirdly, digital health is a complex ecosystem populated by a plethora of stakeholders (e.g., commercial companies, regulators, patients, policymakers, researchers, and healthcare providers) with divergent interests. She concludes that without the establishment of effective governance and ethical oversight the conditions for appropriate data access and use cannot be reached:

> Accountability should not be mechanical checks in an opaque digital ecosystem, rather an opportunity for all stakeholders to embrace responsibility and exercise good judgement. The legal protections afforded to privacy are not intended to be antithetical to the common good of public health, despite how it is often portrayed. The legal basis of public interest for secondary uses of data is an ethically defensible option when robust oversight of data initiatives is in place and transparency is guaranteed. Good governance can strike the right balance between individual protections and promotion of the common good of health.
>
> (Vayena 2021)

Accountability is a long-standing and critical principle in data governance. Addressing accountability in AI, however, requires understanding its broad meaning and consequence. Novelli and others define accountability "as a relation of answerability requiring authority recognition, interrogation and limitation of power" (Novelli et al. 2023, 1881), while criticizing that "accountability in AI is often defined too imprecisely, with undifferentiated references to the values, practices, and measures it encompasses" (ibid, 1871).

In the healthcare and health research sectors, black box approaches, such as deep learning, pose challenges because they cannot explain the decision-making process. This brings up important questions regarding legal accountability and the (oversight of) compliance to ethical standards in AI and calls for rethinking oversight mechanisms in an interdisciplinary fashion (Müller et al. 2021; Holzinger et al. 2025) or with a risk awareness approach (see also chapter "Artificial Intelligence, Sex, and Gender: Hypes, Hopes, and Potential Risks in Biobanking" by Akyüz et al). For health data, the realization of accountability necessitates the review of the core values that are underlying data governance frameworks to ensure that a balance between data privacy and ethical-social considerations is struck and satisfies citizens as well as researchers and innovators (Bak et al. 2022).

4.2 Good Governance as Social License

Establishing a viable governance framework is a challenging yet critical task for any infrastructure, including biobanks. The concept of "good governance" emerged in the late 1980s and was taken up by organizations such as the World Bank or the EU. It covers the rules, practices, and actions that determine how interests are represented, resources are distributed, and authority is exercised in a given context (Landmann 2009). The European Governance White Paper (2001), for instance, focuses on five principles that are critical, namely openness, participation, effectiveness, coherence, and accountability. Whereas it is widely agreed that for biobanks "good governance" is key, balancing, on the one hand, the safeguarding of the rights and expectations of research participants and citizens, and, on the other hand, enabling fair and FAIR access for the scientific community remains a constant challenge (Langhof et al. 2019; Gille et al. 2020; Staunton and de Vries 2020; Abdelhafiz et al. 2022; Akyüz et al. 2024). Robust governance frameworks are said not only to ensure transparency, accountability, and ethical integrity but also to enhance public trust (Gille et al. 2020). Integrating what has been described earlier into a solid governance framework requires the collaboration of a diverse range of actors, practices, and regulatory mechanisms (Kaye 2011).

By establishing clear oversight mechanisms, fostering inclusivity in decision-making, and upholding high ethical standards, staff trainings, etc., biobanks can strengthen their legitimacy and secure their social license to operate. This social license – referring to the public's perception of biobanking as socially acceptable, beneficial, and aligned with societal values – plays a crucial role in maintaining public support and ensuring long-term sustainability (Parsons and Moffat 2014). Put differently, governance can be likened to the construction of a house, where every element of the architectural design must fit in its place to create not only a stable and functional structure but also one that the inhabitants and guests, such as researchers or research participants, find agreeable. Transparency serves as the foundation, ensuring that all practices and procedures are built on trust and accountability and functionally serve those for whom it is built for – much like the architectural axiom of "form follows function".

At this point, I want to emphasize again that in the biobanking field, AI tools are widely regarded with cautious excitement. They are adopted with the expectation to serve as assistive technologies, always operating under human supervision. Autonomous AI systems promise more efficiency, for example, through the assisted writing of draft reports by providing prewritten templates or suggestions. Yet, they also raise concerns about liability, particularly for hospitals, research centers, and software developers, as the rapid pace of AI development increasingly challenges human oversight as the final control mechanism. Especially as the autonomy of AI systems further reinforces their "black box" nature, while poor data quality goes unchallenged due to a lack of sufficient resources. When it concerns human resources, this is often due to a skilled labor shortage with the interdisciplinary expertise and prolonged stamina that is required to oversee complex AI systems effectively and without fatigue (Saenz et al. 2023). In the medical field, explainability and causality play a crucial role – not only for diagnostic and research tasks but also in demonstrating compliance with regulatory requirements (Plass et al. 2023).

To be able to effectively oversee AI systems, it is not necessary to become an expert in every aspect. Rather, digital literacy needs to match the practical realities of patient care (van Voorst 2024), as well as research (Akyüz et al. 2024), or "infrastructuring" (Aarden 2023). Ignoring this could result in the creation of skill sets and experts that are misaligned with practical demands. Accordingly, what van Voorst describes for healthcare providers and patients is equally true for researchers and biobankers:

> Moving forward, it is critical to develop frameworks that not only support the ethical implementation of AI but also safeguard the well-being of both healthcare providers and patients, ensuring that technology enhances rather than complicates care delivery.
>
> (van Voorst 2024)

To achieve this goal, interdisciplinary approaches are required, including human-in-the-loop frameworks and new types of human–AI interfaces, to develop scalable oversight tools that can effectively monitor AI systems, which are operating at unprecedented speeds and volumes that today is managed only insufficiently by either tools or people (Holzinger et al. 2025).

5 Discussion

The interplay between innovation and governance is particularly relevant in an era shaped by the attributes of bigger, better, and faster. Whereas a couple of years ago, size ("critical mass") and standardization were the biggest topics for biobanks, nowadays, the most prevalent challenge is the mere speed by which established norms and technological limits are getting tested, pushed, and redefined. This development is driven by the perceived potential of data-powered, machine-learning systems across the public and private sectors, which influence policy, regulation, and societal norms. The scale (e.g., big data), political momentum (e.g., national strategies), and technological potential (e.g., efficiency) influence policy, regulation, and societal norms – often under the presumption that data are inherently "immaculate" (Bronson 2022). While established norms have always been challenged, the (bio)power that is unleashed through AI systems is undoubtedly productive, yet far from innocent. To exemplify: data may be collected for a specific purpose; this requires a selection that excludes certain data categories – ideally by choice, sometimes by lack of reference data or even ignorance. Alternatively, the absence of data can be mitigated through bias correction methods, which presupposes awareness of the bias in the first place (van Alten et al. 2024). If this is not done, the AI tools cannot yield the desired results.

6 Conclusion

In this chapter, I have drawn the attention to the complex relationship between power, potentiality, and accountability, exemplified by the technology and governance of AI in relation to biobanking, a practice situated at the intersection of healthcare and health research.

While technological innovations have always challenged established norms, the (bio)power they exercise is both productive and deeply consequential. Such advancements not only reshape societal structures, redefine ethical boundaries, and reconfigure economic landscapes but also introduce new forms of control, governance, and inequality. This is exemplified by how AI technologies and regulations influence biobanking – a practice shaped by a society's socio-technical imaginaries, including their societal values and scientific goals.

Disclaimer

Where authors are identified as personnel of the Biobanking and BioMolecular resources Research Infrastructure (BBMRI-ERIC), the authors alone are responsible for the views expressed in this chapter and they do not necessarily represent the decisions, policies, or views of BBMRI-ERIC.

Acknowledgments

Original ideas on rethinking "potentiality" in biobanking date back to my PhD. Over the years, I further benefited from the exchange with interdisciplinary colleagues at BBMRI-ERIC Headquarters, National Nodes, in the context of research projects (most notably INTERVENE/GA 101016775, EuCanImage/GA 952103, Bigpicture/GA 945358, ADOPT BBMRI-ERIC/GA 676550 or EUCAN-Connect/GA 824989), or with representatives from sister research infrastructures – especially patient advocacy and ELSI experts that further inspired this chapter.

Reference List

Aarden, E. 2023. "Infrastructuring European Scientific Integration: Heterogeneous Meanings of the European Biobanking Infrastructure BBMRI-ERIC." *Social Studies of Science* 53 (4): 572–98. https://doi.org/10.1177/03063127231162629.

Abdelhafiz, A. S., M. Ahram, M. E. Ibrahim, A. Elgamri, E. Gamel, R. Labib, and H. Silverman. 2022. "Biobanks in the Low- and Middle-Income Countries of the Arab Middle East Region: Challenges, Ethical Issues, and Governance Arrangements-A Qualitative Study Involving Biobank Managers." *BMC Medical Ethics* 23 (1): 83. https://doi.org/10.1186/s12910-022-00822-8.

Akyüz, Kaya, Mónica Cano Abadía, Melanie Goisauf, and Michaela Th. Mayrhofer. 2024. "Unlocking the Potential of Big Data and AI in Medicine: Insights from Biobanking." *Frontiers in Medicine* 11. https://doi.org/10.3389/fmed.2024.1336588.

Akyuz, K., M. Goisauf, G. M. Martin, M. T. Mayrhofer, S. Antoniou, G. Charalambidou, C. Deltas, A. Malatras, G. Papagregoriou, C. Stefanou, and M. Voutounou. 2024. "Risk Mapping for Better Governance in Biobanking: The Case of Biobank.cy." *Frontiers in Genetics* 15: 1397156. https://doi.org/10.3389/fgene.2024.1397156.

Austin, Claire C. 2020. "The Open Science Ecosystem: A systematic Framework Anchored in Values, Ethics and FAIRER Data." *Available at SSRN*. http://doi.org/10.2139/ssrn.3654298.

Bak, M. A. R., V. I. Madai, M. C. Fritzsche, M. T. Mayrhofer, and S. McLennan. 2022. "You Can't Have AI Both Ways: Balancing Health Data Privacy and Access Fairly." *Frontiers in Genetics* 13: 929453. https://doi.org/10.3389/fgene.2022.929453.

Bak, M. A. R., M. C. Ploem, H. Atesyurek, M. T. Blom, H. L. Tan, and D. L. Willems. 2020. "Stakeholders' Perspectives on the Post-Mortem Use of Genetic and Health-Related Data for Research: A Systematic Review." *European Journal of Human Genetics* 28 (4): 403–16. https://doi.org/10.1038/s41431-019-0503-5.

Bak, M. A. R., M. C. Ploem, H. L. Tan, M. T. Blom, and D. L. Willems. 2023. "Towards Trust-Based Governance of Health Data Research." *Med Health Care Philos* 26 (2): 185–200. https://doi.org/10.1007/s11019-022-10134-8.

Beaulieu, Anne. 2004. "From Brainbank to Database: The Informational Turn in the Study of the Brain." *Studies in History and Philosophy of Biological and Biomedical Sciences* 35 (2): 367–90.

Becker, R., D. Chokoshvili, A. Thorogood, E. S. Dove, F. Molnar-Gabor, A. Ziaka, O. Tzortzatou-Nanopoulou, and G. Comande. 2024. "Purpose Definition as a Crucial Step for Determining the Legal Basis Under the GDPR: Implications for Scientific Research." *Journal of Law and the Biosciences* 11 (1): lsae001. https://doi.org/10.1093/jlb/lsae001.

Beier, Katharina, and Christian Lenk. 2015. "Biobanking Strategies and Regulative Approaches in the EU: Recent Perspectives." *Journal of Biorepository Science for Applied Medicine* 3: 69–81.

Birch, K. 2017. "Rethinking Value in the Bio-economy: Finance, Assetization, and the Management of Value." *Science, Technology, & Human Values* 42: 460–90.

Black, A. 2025. "AI and Democratic Equality: How Surveillance Capitalism and Computational Propaganda Threaten Democracy." In *Bridging the Gap Between AI and Reality*, edited by B. Steffen. Cham: Springer Nature Switzerland, 333–47.

Bronson, Kelly. 2022. *The Immaculate Conception of Data: Agribusiness, Activists, and Their Shared Politics of the Future*. Montreal, Kingston, London and Chicago: McGill-Queen's University Press.

Buyx, A., and B. Prainsack. 2018. "Solidarity Can Make a Difference: Addressing Transformations in Healthcare, Demographics and Technological Replacement." *Bioethics* 32 (9): 537–40. https://doi.org/10.1111/bioe.12540.

Cadigan, R. J., E. Juengst, A. Davis, and G. Henderson. 2014. "Underutilization of Specimens in Biobanks: An Ethical as Well as a Practical Concern?" *Genetics in Medicine* 16 (10): 738–40. https://doi.org/10.1038/gim.2014.38.

Callebaut, W. 2012. "Scientific Perspectivism: A Philosopher of Science's Response to the Challenge of Big Data Biology." *Studies in History and Philosophy of Biological and Biomedical Sciences* 43 (1): 69–80. https://doi.org/10.1016/j.shpsc.2011.10.007.

Cambon-Thomsen, A. 2004. "The Social and Ethical Issues of Post-Genomic Human Biobanks." *Nature Reviews Genetics* 5 (11): 866–73. https://doi.org/10.1038/nrg1473.

Correa, N. K., C. Galvao, J. W. Santos, C. Del Pino, E. P. Pinto, C. Barbosa, D. Massmann, R. Mambrini, L. Galvao, E. Terem, and N. de Oliveira. 2023. "Worldwide AI Ethics: A Review of 200 Guidelines and Recommendations for AI Governance." *Patterns (N Y)* 4 (10): 100857. https://doi.org/10.1016/j.patter.2023.100857.

de Ruijter, A., T. Hervey, and B. Prainsack. 2024. "Solidarity and Trust in European Union Health Governance: Three Ways Forward." *The Lancet Regional Health – Europe* 46: 101047. https://doi.org/10.1016/j.lanepe.2024.101047.

Ducato, Rossana. 2020. "Data Protection, Scientific Research, and the Role of Information." *Computer Law & Security Review* 37: 105412. https://doi.org/10.1016/j.clsr.2020.105412.

European Commission. 2019. *Directorate-General for Communications Networks, Content and Technology. Ethics Guidelines for Trustworthy AI*. Publications Office.

European Commission. 2021a. *Excellence and Trust in Artificial Intelligence*. https://commission.europa.eu/strategy-and-policy/priorities-2019-2024/europe-fit-digital-age/excellence-and-trust-artificial-intelligence_en#:~:text=Benefits%20of%20AI,benefit%20people%2C%20businesses%20and%20governments (14.12.2023).

European Commission. 2021b. White Paper of the European Commission of 12 October 2001 on European Governance COM(2001) 0428 final.

European Commission. 2023. Ethics Advisors and Ethics Advisory Boards. Roles and Function in EU-funded Projects. https://ec.europa.eu/info/funding-tenders/opportunities/docs/2021-2027/horizon/guidance/roles-and-functions-of-ethics-advisory-ethics-advisory-boards-in-ec-funded-projects_he_en.pdf (14.12.2023).

European Data Protection Supervisor. 2020. *A Preliminary Opinion on Data Protection and Scientific Research*. www.edps.europa.eu/sites/default/files/publication/20-01-06_opinion_research_en.pdf (12.02.2023).

Fehr, J., B. Citro, R. Malpani, C. Lippert, and V. I. Madai. 2024. "A Trustworthy AI Reality-Check: The Lack of Transparency of Artificial Intelligence Products in Healthcare." *Frontiers in Digital Health* 6: 1267290. https://doi.org/10.3389/fdgth.2024.1267290.

Foucault, Michel, and Paul Rabinow. 1984. *The Foucault Reader*. 1st ed. New York: Pantheon Books.

Foucault, Michel, and Paul Rabinow. 2002. *The Essential Works of Michel Foucault, 1954–1984*. London: Penguin.

Gille, F., E. Vayena, and A. Blasimme. 2020. "Future-Proofing Biobanks' Governance." *European Journal of Human Genetics* 28 (8): 989–96. https://doi.org/10.1038/s41431-020-0646-4.

Goisauf, M., and M. Cano Abadia. 2022. "Ethics of AI in Radiology: A Review of Ethical and Societal Implications." *Frontiers in Big Data* 5: 850383. https://doi.org/10.3389/fdata.2022.850383.

Haeck, Pieter. 2025. "Trump's $500B AI Plan Is 'Slap' in the Face for Europe." *Politico*. https://www.politico.eu/article/us-500-ai-europe-donald-trump-global-leadership-eu-social-media-china/ (09.06.2025).

Hagen, H. E., and J. Carlstedt-Duke. 2004. "Building Global Networks for Human Diseases: Genes and Populations." *Nature Medicine* 10 (7): 665–7. https://doi.org/10.1038/nm0704-665.

Hämäläinen, I., O. Törnwall, B. Simell, K. Zatloukal, M. Perola, and G. B. van Ommen. 2019. "Role of Academic Biobanks in Public-Private Partnerships in the European Biobanking and BioMolecular Resources Research Infrastructure Community." *Biopreserv Biobank* 17 (1): 46–51. https://doi.org/10.1089/bio.2018.0024.

Haraway, D. J. 1988. *Situated Knowledges: The Science Question in Feminism and the Privilege of Partial Perspective*.

Hilgartner, Stephen. 1995. "Biomolecular Databases: New Communication Regimes for Biology?" *Science Communication* 17 (2): 240–63.

Hoeyer, K. 2019. "Data as Promise: Reconfiguring Danish Public Health Through Personalized Medicine." *Social Studies of Science* 49 (4): 531–55. https://doi.org/10.1177/0306312719858697.

Holub, Petr, Florian Kohlmayer, Fabian Prasser, Michaela Th Mayrhofer, Irene Schlünder, Gillian M. Martin, Sara Casati, Lefteris Koumakis, Andrea Wutte, Łukasz Kozera, Dominik Strapagiel, Gabriele Anton, Gianluigi Zanetti, Osman Ugur Sezerman, Maimuna Mendy, Dalibor Valík, Marialuisa Lavitrano, Georges Dagher, Kurt Zatloukal, GertJan B. van Ommen, and Jan-Eric Litton. 2018. "Enhancing Reuse of Data and Biological Material in Medical Research: From FAIR to FAIR-Health." *Biopreservation and Biobanking* 16 (2): 97–105. https://doi.org/10.1089/bio.2017.0110.

Holzinger, A., K. Zatloukal, and H. Muller. 2025. "Is Human Oversight to AI Systems Still Possible?" *New Biotechnology* 85: 59–62. https://doi.org/10.1016/j.nbt.2024.12.003.

Hummel, P., and M. Braun. 2020. "Just Data? Solidarity and Justice in Data-Driven Medicine." *Life Sciences, Society and Policy* 16 (1): 8. https://doi.org/10.1186/s40504-020-00101-7.
Jasanoff, Sheila, and Sang-Hyun Kim. 2015. *Dreamscapes of Modernity: Sociotechnical Imaginaries and the Fabrication of Power*. 1st ed. Chicago: University of Chicago Press.
Kaye, J. 2011. "From Single Biobanks to International Networks: Developing E-Governance." *Human Genetics* 130 (3): 377–82. https://doi.org/10.1007/s00439-011-1063-0.
KPMG Enterprise. 2018. *Venture Pulse: Q1 '18 Global Analysis of Venture Funding*. https://assets.kpmg.com/content/dam/kpmg/xx/pdf/2018/04/kpmg-venture-pulse-q1-2018.pdf (09.06.2025).
Landecker, Hannah. 2007. *Culturing Life: How Cells Became Technologies*. Cambridge, MA: Harvard University Press.
Landmann, Todd. 2009. *Concepts Matter: Delineating Democracy, Governance and Human Rights*. https://www.idea.int/sites/default/files/publications/chapters/the-role-of-the-european-union-in-democracy-building/eu-democracy-building-discussion-paper-51.pdf (09.06.2025).
Langhof, H., J. Schwietering, and D. Strech. 2019. "Practice Evaluation of Biobank Ethics and Governance: Current Needs and Future Perspectives." *Journal of Medical Genetics* 56 (3): 176–85. https://doi.org/10.1136/jmedgenet-2018-105617.
Lekadir, K., A. F. Frangi, A. R. Porras, B. Glocker, C. Cintas, C. P. Langlotz, E. Weicken, F. W. Asselbergs, F. Prior, G. S. Collins, G. Kaissis, G. Tsakou, I. Buvat, J. Kalpathy-Cramer, J. Mongan, J. A. Schnabel, K. Kushibar, K. Riklund, K. Marias, L. M. Amugongo, L. A. Fromont, L. Maier-Hein, L. Cerda-Alberich, L. Marti-Bonmati, M. J. Cardoso, M. Bobowicz, M. Shabani, M. Tsiknakis, M. A. Zuluaga, M. C. Fritzsche, M. Camacho, M. G. Linguraru, M. Wenzel, M. De Bruijne, M. G. Tolsgaard, M. Goisauf, M. Cano Abadia, N. Papanikolaou, N. Lazrak, O. Pujol, R. Osuala, S. Napel, S. Colantonio, S. Joshi, S. Klein, S. Ausso, W. A. Rogers, Z. Salahuddin, M. P. A. Starmans, and Future-Ai Consortium. 2025. "FUTURE-AI: International Consensus Guideline for Trustworthy and Deployable Artificial Intelligence in Healthcare." *BMJ* 388: e081554. https://doi.org/10.1136/bmj-2024-081554.
Lenoir, Timothy. 2002. "Science and the Academy of the 21st Century: Does Their Past Have a Future in an Age of Computer-Mediated Networks?" In *Ideale Akademie: Vergangene Zukunft oder konkrete Utopie?*, edited by Wilhelm Voßkamp. Berlin: Forschungsberichte der Interdisziplinären Arbeitsgruppen der Berlin-Brandenburgischen Akademie der Wissenschaften, 113–29.
Mayrhofer, M. T. 2013a. "About the New Significance and the Contingent Meaning of Biological Material and Data in Biobanks." *History & Philosophy of the Life Sciences* 35 (3): 449–67.
Mayrhofer, M. T. 2013b. "Biobanks and Research Infrastructures." In *Strengthening Intangible Infrastructures*, edited by Elisabeth Kampferer, Andreas Koch and Clemens Sedmak, 287–300. Cambridge Scholards Publishing.
Mittelstadt, B. D. 2021. "Near-Term Ethical Challenges of Digital Twins." *Journal of Medical Ethics*. https://doi.org/10.1136/medethics-2021-107449.
Mittelstadt, B. D., and L. Floridi. 2016. "The Ethics of Big Data: Current and Foreseeable Issues in Biomedical Contexts." *Science and Engineering Ethics* 22 (2): 303–41. https://doi.org/10.1007/s11948-015-9652-2.
Müller, Heimo, Michaela Mayrhofer, Evert-Ben Veen, and Andreas Holzinger. 2021. "The Ten Commandments of Ethical Medical AI." *Computer* 54: 119–23. https://doi.org/10.1109/mc.2021.3074263.
Murtagh, M. J., M. Machirori, C. L. Gaff, M. T. Blell, J. de Vries, M. Doerr, E. S. Dove, A. Duncanson, J. Hastings Ward, R. Hendricks-Sturrup, C. W. L. Ho, A. Johns, Y. Joly, K. Kato, K. Katsui, J. Kumuthini, F. Maleady-Crowe, A. Middleton, R. Milne, J. T. Minion, M. Matshaba, S. Mulrine, C. Patch, R. Ryan, and W. Viney. 2021. "Engaged Genomic Science Produces Better and Fairer Outcomes: An Engagement Framework for Engaging and Involving Participants, Patients and Publics in Genomics Research and

Healthcare Implementation." *Wellcome Open Research* 6: 311. https://doi.org/10.12688/wellcomeopenres.17233.1.

Myong, Y., D. Yoon, B. S. Kim, Y. G. Kim, Y. Sim, S. Lee, J. Yoon, M. Cho, and S. Kim. 2023. "Evaluating Diagnostic Content of AI-Generated Chest Radiography: A Multi-Center Visual Turing Test." *PLoS ONE* 18 (4): e0279349. https://doi.org/10.1371/journal.pone.0279349.

Novelli, C., M. Daddeo, and L. Floridi. 2023. "Accountability in Artificial Intelligence: What It Is and How It Works." *AI & Society* 39: 1871–82. https://doi.org/10.1007/s00146-023-01635-y.

OECD. 2024. "AI Principles." *Adopted in 2019*. https://oecd.ai/en/ai-principles (10.12.2024).

Okaibedi Eke, Damian, Kutoma Wakunuma, Simisola Akintoye, and George Ogoh. 2025. *Trustworthy AI: African Perspective*. Cham: Palgrave Macmillan.

Oldoni, Emanuela, Alain van Gool, Laura García Bermejo, Andreas Scherer, Michaela Th Mayrhofer, Francesco Florindi, Jacques Demotes, Christine Kubiak, Anne-Charlotte Fauvel, Florence Bietrix, Anton Ussi, and Antonio L. Andreu. 2021. "Biomarker Research and Development for Coronavirus Disease 2019 (COVID-19): European Medical Research Infrastructures Call for Global Coordination." *Clinical Infectious Diseases* 72 (10): 1838–42. https://doi.org/10.1093/cid/ciaa1250.

Pandya, Jayshree 2019. "Biobanking Is Changing The World." *Forbes*. https://www.forbes.com/sites/cognitiveworld/2019/08/12/biobanking-is-changing-the-world/ (09.06.2025).

Park, Alice. 2009. "Biobanks – 10 Ideas Changing the World Right Now." *Time Magazine*. https://content.time.com/time/specials/packages/article/0,28804,1884779_1884782_1884766,00.html (09.06.2025).

Parsons, Richard, and Kieren Moffat. 2014. "Constructing the Meaning of Social Licence." *Social Epistemology* 28 (3–4): 340–63. https://doi.org/10.1080/02691728.2014.922645.

Peloquin, D., M. DiMaio, B. Bierer, and M. Barnes. 2020. "Disruptive and Avoidable: GDPR Challenges to Secondary Research Uses of Data." *European Journal of Human Genetics* 28 (6): 697–705. https://doi.org/10.1038/s41431-020-0596-x.

Plass, M., M. Kargl, T. R. Kiehl, P. Regitnig, C. Geissler, T. Evans, N. Zerbe, R. Carvalho, A. Holzinger, and H. Muller. 2023. "Explainability and Causability in Digital Pathology." *The Journal of Pathology: Clinical Research* 9 (4): 251–60. https://doi.org/10.1002/cjp2.322.

Prainsack, Barbara. 2017. *Personalized Medicine: Empowered Patients in the 21st Century, Biopolitics: Medicine, Technoscience, and Health in the 21st Century*. New York: New York University Press.

Radclyffe, C., M. Ribeiro, and R. H. Wortham. 2023. "The Assessment List for Trustworthy Artificial Intelligence: A Review and Recommendations." *Frontiers in Artificial Intelligence* 6: 1020592. https://doi.org/10.3389/frai.2023.1020592.

Rajpurkar, P., J. Irvin, R. L. Ball, K. Zhu, B. Yang, H. Mehta, T. Duan, D. Ding, A. Bagul, C. P. Langlotz, B. N. Patel, K. W. Yeom, K. Shpanskaya, F. G. Blankenberg, J. Seekins, T. J. Amrhein, D. A. Mong, S. S. Halabi, E. J. Zucker, A. Y. Ng, and M. P. Lungren. 2018. "Deep Learning for Chest Radiograph Diagnosis: A Retrospective Comparison of the CheXNeXt Algorithm to Practicing Radiologists." *PLoS Medicine* 15 (11): e1002686. https://doi.org/10.1371/journal.pmed.1002686.

Rheinberger, Hans-Jörg. 1997. *Toward a History of Epistemic Things: Synthesizing Proteins in the Test Tube, Writing Science*. Stanford, CA: Stanford University Press.

Ribeiro Junior, H. L. 2024. "AI Ethics in Medical Research: The 2024 Declaration of Helsinki." *Lancet* 404 (10467): 2048–9. https://doi.org/10.1016/S0140-6736(24)02376-6.

Rush, A., D. R. Catchpoole, R. Ling, A. Searles, P. H. Watson, and J. A. Byrne. 2020. "Improving Academic Biobank Value and Sustainability Through an Outputs Focus." *Value Health* 23 (8): 1072–8. https://doi.org/10.1016/j.jval.2020.05.010.

Russo, Lucia, and Noah Oder. 2023. "How Countries Are Implementing the OECD Principles for Trustworthy AI." *OECD.AI Observatory*. https://oecd.ai/en/wonk/national-policies-2 (09.06.2025).

Saenz, A. D., Z. Harned, O. Banerjee, M. D. Abramoff, and P. Rajpurkar. 2023. "Autonomous AI Systems in the Face of Liability, Regulations and Costs." *NPJ Digital Medicine* 6 (1): 185. https://doi.org/10.1038/s41746-023-00929-1.

Slokenberga, S., O. Tzortzatou, and J. Reichel, eds. 2021. *GDPR and Biobanking Individual Rights, Public Interest and Research Regulation across Europe, Law, Governance and Technology Series*. Springer International Publishing.

Stargardter, Gabriel. 2025. "Details of 110 Billion Euros in Investment Pledges at France's AI Summit." *Reuters*. www.reuters.com/technology/artificial-intelligence/details-110-billion-euros-investment-pledges-frances-ai-summit-2025-02-10/ (28.02.2025).

Staunton, C., and J. de Vries. 2020. "The Governance of Genomic Biobank Research in Africa: Reframing the Regulatory Tilt." *The Journal of Law and the Biosciences* 7 (1): lsz018. https://doi.org/10.1093/jlb/lsz018.

Teng, Tina. 2025. "Chinese Startup DeepSeek Rattles Global Markets as Nvidia Shares Plunge." *Euronews*. www.euronews.com/business/2025/01/28/chinese-startup-deepseek-rattles-global-markets-as-nvidia-shares-plunge (05.02.2025).

van Alten, S., B. W. Domingue, J. Faul, T. Galama, and A. T. Marees. 2024. "Reweighting UK Biobank Corrects for Pervasive Selection Bias Due to Volunteering." *International Journal of Epidemiology* 53 (3). https://doi.org/10.1093/ije/dyae054.

van Voorst, Roanne. 2024. "Challenges and Limitations of Human Oversight in Ethical Artificial Intelligence Implementation in Health Care: Balancing Digital Literacy and Professional Strain." *Mayo Clinic Proceedings: Digital Health* 2 (4): 559–63. https://doi.org/10.1016/j.mcpdig.2024.08.004.

Vayena, E. 2021. "Value from Health Data: European Opportunity to Catalyse Progress in Digital Health." *Lancet* 397 (10275): 652–3. https://doi.org/10.1016/S0140-6736(21)00203-8.

Vetter, Dennis, Julia Amann, Frédérick Bruneault, Megan Coffee, Boris Düdder, Alessio Gallucci, Thomas Krendl Gilbert, Thilo Hagendorff, Irmhild van Halem, Eleanore Hickman, Elisabeth Hildt, Sune Holm, Georgios Kararigas, Pedro Kringen, Vince I. Madai, Emilie Wiinblad Mathez, Jesmin Jahan Tithi, Magnus Westerlund, Renee Wurth, Roberto V. Zicari, and Z. Inspection® initiative. 2023. "Lessons Learned from Assessing Trustworthy AI in Practice." *Digital Society* 2 (3): 35. https://doi.org/10.1007/s44206-023-00063-1.

Waldby, C. 2002. "Stem Cells, Tissue Cultures and the Production of Biovalue." *Interdisciplinary Journal for the Social Study of Health, Illness and Medicine* 6: 305–23.

Waldby, C. and R. Mitchell. 2006. *Tissue Economies: Blood, Organs and Cell Lines in Late Capitalism*. Durham: Duke University Press.

Wilkinson, M. D., M. Dumontier, I. J. Aalbersberg, G. Appleton, M. Axton, A. Baak, N. Blomberg, J. W. Boiten, L. B. da Silva Santos, P. E. Bourne, J. Bouwman, A. J. Brookes, T. Clark, M. Crosas, I. Dillo, O. Dumon, S. Edmunds, C. T. Evelo, R. Finkers, A. Gonzalez-Beltran, A. J. Gray, P. Groth, C. Goble, J. S. Grethe, J. Heringa, P. A. t Hoen, R. Hooft, T. Kuhn, R. Kok, J. Kok, S. J. Lusher, M. E. Martone, A. Mons, A. L. Packer, B. Persson, P. Rocca-Serra, M. Roos, R. van Schaik, S. A. Sansone, E. Schultes, T. Sengstag, T. Slater, G. Strawn, M. A. Swertz, M. Thompson, J. van der Lei, E. van Mulligen, J. Velterop, A. Waagmeester, P. Wittenburg, K. Wolstencroft, J. Zhao, and B. Mons. 2016. "The FAIR Guiding Principles for Scientific Data Management and Stewardship." *Scientific Data* 3: 160018. https://doi.org/10.1038/sdata.2016.18.

World Medical Association. 2024. "WMA Declaration of Helsinki – Ethical Principles for Medical Research Involving Human Participants." www.wma.net/policies-post/wma-declaration-of-helsinki/ (10.12.2024).

Zhang, J., and Z. M. Zhang. 2023. "Ethics and Governance of Trustworthy Medical Artificial Intelligence." *BMC Medical Informatics and Decision Making* 23 (1): 7. https://doi.org/10.1186/s12911-023-02103-9.

Zuboff, S. 2020. *The Age of Surveillance Capitalism: the Fight for a Human Future at the New Frontier of Power*. New York: PublicAffairs.

Legal Sources

Charter of Fundamental Rights of the European Union 2012/C 326/02 OJ C 326.
Council of Europe. 1997. Convention for the Protection of Human Rights and Dignity of the Human Being with Regard to the Application of Biology and Medicine: Convention on Human Rights and Biomedicine (Oviedo Convention). www.coe.int/en/web/human-rights-and-biomedicine/oviedo-convention (10.02.2025).
Regulation (EU) 2024/1689 Laying Down Harmonised Rules on Artificial Intelligence and Amending Regulations (EC) No 300/2008, (EU) No 167/2013, (EU) No 168/2013, (EU) 2018/858, (EU) 2018/1139 and (EU) 2019/2144 and Directives 2014/90/EU, (EU) 2016/797 and (EU) 2020/1828 (Artificial Intelligence Act) [2024] OJ L.

Conclusion

12 Learnings From Unboxing the Black Box

Michaela Th. Mayrhofer, Santa Slokenberga, and Signe Mežinska

Biobanks are at the forefront of AI-driven transformations in the life sciences and medicine, where health research and healthcare intersect. AI-powered tools promise to reshape how diseases are diagnosed, treated, and even prevented, proposing a more efficient, data-driven personalized medicine powered by big data research. Whereas in health research, machine learning algorithms can, for instance, identify complex patterns in clinical data and uncover new risk factors, AI applications can accelerate diagnosis through computer-aided imaging in healthcare.

In both healthcare and health research, AI's effectiveness is contingent on accessing and using sensitive information from individuals or groups. Depending on the context, these individuals are called patients (e.g., in clinical settings), research participants (e.g., in clinical trials), data subjects (e.g., in data protection regulations), or citizens (e.g., in discussions of personal choice). Assembled in vast datasets, AI tools harvest genomic, health, clinical, environmental, and lifestyle data. AI-driven systems are data-hungry and require ever larger and more detailed datasets to refine their algorithms.

In 11 chapters, this edited volume examines key ethical, legal, and societal aspects of AI in the context of biobanking. By looking at liability aspects, the precautionary principle, risk assessment, principles of accountability, and drawing attention to trustworthy governance, it showcases how AI systems are used in biobanking and explores what is necessary to retain social acceptance.

Biobanks, we argue, have both the experience and awareness necessary for handling sensitive data and careful application of new technologies, such as AI. Biobanks, for instance, rely on institutionalized safeguards such as the collaboration with research ethics committees (RECs) that assess the ethical soundness of research projects prior to their execution and can remain doing so as long as expertise for assessing the use of new technologies is warranted (see chapter "Biobank-Based Research Employing AI Techniques: Challenges for Research Ethics Committees" by Signe Mežinska). Moreover, we showcase that clear procedures, good communication, and a solid governance are essential requirements to employ AI systems, so that they are both trustworthy and socially accepted by multiple stakeholders (see chapters "Social Acceptance of Artificial Intelligence in Biobanking" by Caboux et al. and "Unique by Design – the Need for Individualized Models of Trust in Biobanking AI" by Anna Clareborn). This requires

DOI: 10.4324/9781032634487-18

a constant dialogue, which can be assisted by AI tools if the purpose they serve is clearly defined and communicated (see chapter "Artificial Intelligence and Communication With Research Participants in the Process of Biobanking Human Biological Material" by Jarosław Sak and Jakub Pawlikowski). We further claim that AI usage can be made explainable by honestly addressing transparency concerns (see chapter "Tackling AI Transparency Concerns in Biomedical Research: Bringing a Communication-Participatory Approach to the Conversation" by Leslie Salgado Arzuaga). However, at the same time, we are also aware of the possible costs of transparency and the need for further inquiries as science and technology evolve and assessment of the struck balances. Transparency is a reachable goal, if compliance procedures integrate the precautionary principle (see chapter "A Possible Risk Governance Approach for AI in Health Research and Biobanking" by Ilaria Colussi) and ensure fairness (see chapter "Fairness of an AI System in the Case of a Biobank of Images and Imaging Biomarkers" by Valentina Colcelli).

All this presupposes that AI systems are trained on truly representative datasets, as otherwise they may unintentionally reinforce health disparities, disproportionately affect and (re)create vulnerabilities, and fail to deliver improved healthcare (see chapter "Powerful AI: Between Accountability and Potentiality" by Michaela Th. Mayrhofer). Otherwise, such disparities not only undermine trust in AI-driven healthcare solutions but also increase legal and ethical liabilities for developers and healthcare providers (see chapter "Liability for Damages, AI, and Machine Learning for Digital Pathology as New Challenges for Biobanks" by Dorota Krekora-Zając). Knowing about gaps in data requires taking measures by law (see chapter "The Approach of EU Law to Tackling Statistical Bias in Medicine" by Katharina Ó Cathaoir) and overcoming bias in practice (see chapter "Artificial Intelligence, Sex, and Gender: Hypes, Hopes, and Potential Risks in Biobanking" by Akyüz et al.).

AI is revolutionizing multiple aspects of society and one of the most profound transformations is taking place in the life sciences and medicine at the intersection of health research and healthcare as exemplified on biobanking. Because of their experience in handling sensitive data and integrating new technologies, we demonstrated several long-standing and new ethical, legal, and policy instruments that can shed light on the "black box" that is algorithmic intelligence. The European Union's Artificial Intelligence Act,[1] for instance, adopts a risk-based approach, categorizing AI tools by their impact on individuals and society. This far-reaching legislation applies to developers, deployers, and users, requiring compliance from both public and private actors. However, at the same time, its relevance at the research stage depends on several considerations. Alongside regulations like the European Health Data Space,[2] the EU seeks to balance innovation with ethical principles such as autonomy, dignity, and privacy, while mitigating AI-related risks.

For biobanks, however, the integration and regulatory compliance of AI systems is only one element of many to consider for the operation of an infrastructure for the life sciences and medicine. Rather, the direct relevance for biobanks is limited. Indeed, AI regulation coexists with several ethical and legal requirements, and making sense of them in practice remains a challenging task that requires multiple disciplines. To guarantee the fair use of and access to the resources managed in and by public biobanks for the benefit of all members of society, interdisciplinary collaboration is essential. In a first attempt, our book brought together experts from law, philosophy, political science, communication, and social sciences. However, it requires more, including experts from technology, medicine, healthcare, patient advocacy, and computer science, among others. Ultimately, it is a joint effort to address challenges related to individual and societal benefits, bias, transparency, and accountability that new technologies, such as AI-driven systems, impose and it is our task to shed light on the road ahead so that we do not stumble.

Notes

1 AI Act: https://eur-lex.europa.eu/eli/reg/2024/1689/oj/eng (23.03.2025).
2 EHDS: https://eur-lex.europa.eu/eli/reg/2025/327/oj/eng (21.03.2025).

Index

Note: Page numbers in *italics* indicate a figure, and page numbers in **bold** indicate a table on the corresponding page.